Water Science Reviews 1

Water Science Reviews 1

EDITED BY

FELIX FRANKS

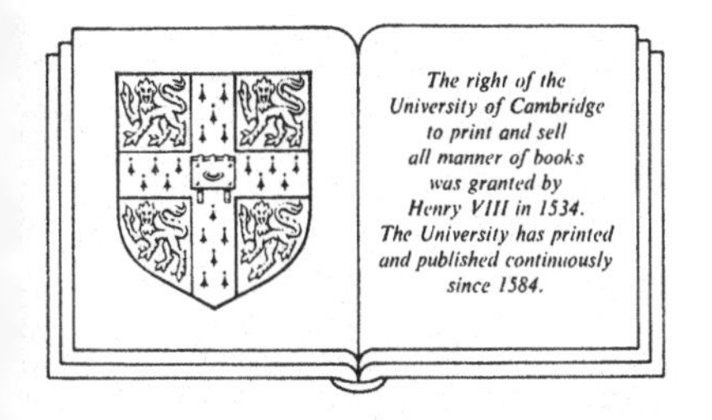

CAMBRIDGE UNIVERSITY PRESS

CAMBRIDGE

LONDON NEW YORK NEW ROCHELLE

MELBOURNE SYDNEY

CAMBRIDGE UNIVERSITY PRESS
Cambridge, New York, Melbourne, Madrid, Cape Town, Singapore, São Paulo, Delhi

Cambridge University Press
The Edinburgh Building, Cambridge CB2 8RU, UK

Published in the United States of America by Cambridge University Press, New York

www.cambridge.org
Information on this title: www.cambridge.org/9780521371728

First published 1985
This digitally printed version 2008

A catalogue record for this publication is available from the British Library

ISBN 978-0-521-37172-8 hardback
ISBN 978-0-521-09902-8 paperback

Contents

Preface

Water is probably the most eccentric chemical known to man. It is the only substance which exists in all three states of matter on this planet and it is also the only inorganic liquid which occurs naturally. On the other hand it is the natural substrate for all *in vivo* processes and the lifelong environment for many species.

It has long been a source of wonder for philosophers, painters, poets, composers and, much more recently, for physicists, chemists, biologists, and even astronomers. Although there is now general agreement that its remarkable properties derive from hydrogen bonding, we are still at a loss how to explain the bulk physical properties in terms of the molecular structure of the H_2O molecule or, indeed, the water dimer. Intensive study of water dates from the 1960s, after the foundation had been laid by Bernal and Fowler in their classic paper of 1933. The realization that water plays a central role in maintaining native biopolymer structures was slow in coming, but since 1970 the importance of hydration figures largely in the protein literature.

Over a period of thirteen years I was involved in the publication of the seven-volume work *Water – A Comprehensive Treatise*. For various reasons both of a personal and practical nature, I decided not to continue with this project. On the other hand, there are still many topics where water takes a central position and which are due for a review. There are also other topics which featured in the *Comprehensive Treatise* but where recent progress has been so rapid that an update is opportune. Three such topics feature in Volume 1 of *Water Science Reviews*.

The reviews are aimed not only at the specialist who finds it difficult to keep up to date with original research beyond a very limited range but also at those interested in a wide range of basic and applied research on water which may extend far beyond their primary field. The reports are to be concise and critical in the well-tried manner of the *Comprehensive Treatise*, with a common focus on the central position adopted by water in the systems and processes covered.

It is a pleasure to acknowledge my gratitude to all my former collaborators, and I hope that *Water Science Reviews* will continue the task of convincing the scientific community at large of the universal importance of water. I would be particularly grateful for suggestions (topics and suitable authors) of aspects of water science which might feature in future volumes of this journal.

Cambridge, May 1985 FELIX FRANKS

Structural Studies of Water by Neutron Diffraction

J. C. DORE

Physics Laboratory, University of Kent, Canterbury, Kent

1. Introduction

Water is a common substance. Its basic properties are widely known yet it represents a considerable challenge to the scientist who wishes to understand its behaviour on a molecular scale. The apparent familiarity with water in its bulk liquid phase creates a deceptive illusion about the simplicity of the molecular interactions which govern these microscopic properties since it is found that water is an extremely complex material. Although a considerable amount of information has been gathered over several decades of research investigation and presented in numerous individual reviews a clear picture of the detailed behaviour has not yet been unambiguously determined. In this context the interest remains as strong as ever and the development of new techniques for putting together the 'final' pieces of the jigsaw remains as a tantalizing challenge to a wide range of the scientific community. The series of articles comprising *Water: A Comprehensive Treatise* [**1**] provides a clear indication of the way the subject has developed in recent years, but it is also apparent that some of the work described in the earlier volumes has now been superseded by new measurements and new ideas.

In this article, the most recent developments in neutron diffraction techniques will be described. The wide range of experiments undertaken by various research groups in many countries has been initiated in an attempt to provide a more complete picture of the spatial correlations that exist between molecules in the liquid. The overriding feature affecting the interaction between water molecules in the condensed state centres on the phenomenon of hydrogen-bonding, in which strongly orientation-dependent forces are known to influence the structural configuration. The exact nature of the time-averaged molecular correlations remains to be established. The most obvious and direct way of observing this 'structure' is by means of diffraction measurements in which the intensity of coherently scattered radiation from the sample material is detected in some suitable diffractometer or spectrometer. The essential information about the spatial distribution of the scattering is therefore contained in the phase differences of the scattered waves which lead to interference effects in the observed diffraction data. Since the characteristic dimensions associated with structure on the atomic or molecular scale are in the region of 1–10 Å, it follows that diffraction measurements are most conveniently made with incident radiation which has a wavelength in the 0.1 to 2 Å range.

The use of X-ray methods for structural studies of liquids is well established and has a long history, but the use of thermal neutrons is a relatively recent advance which has developed rapidly over the last fifteen years. This advance has been achieved by the construction of high-flux reactors specifically designed for neutron research purposes and the associated building of specialized instruments to obtain the required experimental information. The facilities available to scientists engaged in this work have continued to improve with gains from higher neutron fluxes, more efficient detectors and improved data collection systems. The impact of this technology on the scientific study of the structural and dynamic properties of condensed materials has been immense. In the following sections an attempt will be made to show how these new methods have been used to study the structural properties of water. It will be seen that, despite considerable improvement in available knowledge, information and understanding, the process is by no means finished and further experimental study will be needed to resolve many of the questions which still remain. The enigma of 'water' continues.

2. Diffraction Methods

2.1. *Basic Features of Neutron Scattering*

The theoretical formalism for the scattering of X-rays and thermal neutrons by molecular liquids has been well described elsewhere [**2**, **3**], and will not be repeated here. The essential features of the observed diffraction pattern can be directly related to a structure factor which characterizes the basic arrangement of scattering centres for the sample under study. In ordered materials, such as crystals, the regular arrangement of the lattice leads to a series of well-defined peaks governed by the Laue conditions. In the case of partially disordered materials such as liquids and amorphous solids the observed pattern is composed of a broader oscillatory structure but still contains structural information through the phase differences arising in the interference of the scattered waves.

Thermal neutrons provide a convenient probe for structural studies of many materials; a neutron with a de Broglie wavelength of 1 Å has an energy of 82.8 meV and a reciprocal velocity of 252 μs m^{-1}. Unlike X-rays which are scattered by electromagnetic interaction with the atomic electron distribution, the neutron is scattered by the nucleus which behaves as a point scatterer since its size ($\sim 10^{-15}$ m) is much smaller than the wavelength of the incident neutron ($\sim 10^{-10}$ m). In this respect, the formalism for describing the scattering of neutrons is much simpler than that for X-rays since there is no need to introduce an atomic scattering factor to represent the spatial distribution for each individual scattering unit. There are, however, other characteristics arising from the nuclear interaction mechanism which must be taken into account and have an important bearing on the conduct of the experiments and the interpretation of the results.

For low energies, the nuclear scattering is isotropic (S-wave scattering) and for a scattering centre of zero spin it is entirely coherent. The scattering amplitude is therefore represented in terms of a single value, b, for the bound-atom coherent scattering length, provided the neutron energy is not close to a compound nucleus resonance. For nuclei with spin, there are two terms, b_+ and b_- for parallel and anti-parallel alignment of the spins of the neutron and the nucleus. Since there is normally no preferential spin orientation, a statistical average may be taken which results in two components that characterize the effective scattering. The coherent scattering length b_{coh} again defines that part of the scattered wave which conveys information on structure through the phase factors, but the additional component due to spin incoherence does not exhibit interference effects and therefore cannot yield any structural information. Furthermore, the b_{coh}-values for different isotopes will not, in general, be the same and there is therefore a further contribution from isotopic incoherence since there is normally no preferential ordering of the isotopes in the sample. [4] The importance of the incoherent contribution is particularly marked for hydrogen-containing materials and will be discussed more fully in section 6. Tables of neutron scattering values for both natural elements and isotopes are presented in various textbooks dealing with neutron scattering; the most recent compilation is due to Sears [5], although an earlier version due to Koester [6] is frequently used.

2.2. *The Description of Liquid Structure*

For atomic liquids it is usual to define a liquid structure factor $S(Q)$ by the relation

$$S(Q) = 1+\rho\int \exp(\mathrm{i}\,\mathbf{Q}\cdot\mathbf{r})[g(\mathbf{r})-1]\,\mathrm{d}^3\mathbf{r}, \tag{2.1}$$

where $\mathbf{Q}$ (or $\mathbf{k}$) is the scattering vector, ρ is the atomic number density and $g(\mathbf{r})$ is the atomic pair correlation function which represents the probability of finding a pair of atoms with a separation $\mathbf{r}$. The inverse transform relation may be written as:

$$g(\mathbf{r}) = 1+\frac{1}{(2\pi)^3\rho}\int \exp(-\mathrm{i}\,\mathbf{Q}\cdot\mathbf{r})[S(Q)-1]\,\mathrm{d}^3\mathbf{Q}. \tag{2.2}$$

In the case of molecular liquids these expressions may be generalized, but the loss of spherical symmetry introduces an additional complexity into the description. Two equivalent but alternative methods may be used in which different properties of the system are emphasized. The spatial correlation

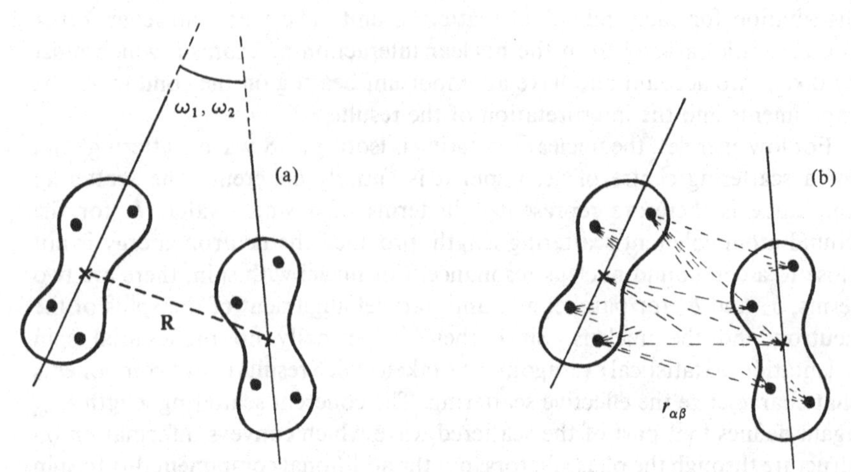

Figure 1. The formation for representation of spatial distributions in molecular liquids: (a) $G(R, \omega_1\omega_2)$; (b) $g_{\alpha\beta}(r)$.

between two identical molecules is represented schematically in figure 1(a). The full pair correlation function may now be written as $G(\mathbf{R}, \omega_1\omega_2)$, where $\mathbf{R}$ represents the distance between the molecular centres and ω_1, ω_2 are the Euler angles corresponding to the orientation of molecular axes. This function gives a complete description of the system but consists of multiple variables and is not accessible to direct evaluation from diffraction observations. For individual molecules, the general expression may be simplified by symmetry requirements, but further approximations are usually required and the formula must be written as an expansion in terms of spherical harmonics. This formalism has been used by Zeidler and co-workers [7] for the treatment of data for small molecules such as chloroform and emphasizes the orientational correlations between adjacent molecules. The method is widely adopted for studies of the dynamic properties of liquids where re-orientational processes play a major role in the time-dependent behaviour. However, its use as a suitable representation for the static structural properties is severely limited because the series is found to be slowly convergent and it is difficult to obtain satisfactory information over a wide range of R-values. Although this type of description may be required for large molecules, it seems more convenient to use a formalism based on the partial pair correlation functions dependent on the spatial relationship of individual atoms as shown in figure 1(b). In this case, the atoms of each molecule are separately labelled and lead to a series of partial pair-correlation functions corresponding to the spatial distribution for different atomic sites. For water, with two components, there are therefore three independent pair correlation functions $g_{OO}(r)$, $g_{OH}(r)$ and $g_{HH}(r)$. Representation of the liquid structure by this means emphasizes the separation of distances within the same molecule and between adjacent molecules but contains no specific information about the relative orientation of neighbouring molecules. It is therefore, in certain respects, an incomplete description of the system, although some information on the orientational correlations can often be deduced from peak positions in the partial $g_{\alpha\beta}(r)$ curves. The incomplete nature of the description is clearly illustrated in the case of homonuclear diatomic molecules such as N_2 since there can be only one pair-correlation function $g_{NN}(r)$ which must be representative of the more complex function $G(\mathbf{R},\omega_1\,\omega_2)$ containing details of the orientational correlations. This important feature of diffraction studies on disordered materials is often overlooked, since the information is of a fundamentally different form from that arising in crystallography where a complete description of the structural properties can be confidently established from only a few independent experiments. The main aim of diffraction experiments on liquids is therefore to obtain a precise evaluation of the partial $g_{\alpha\beta}(r)$ functions which characterize the material under study and to use modelling or computer simulation to extract additional details which are not directly accessible to experimental investigation.

2.3. *The Molecular Structure Factor*

The generalization of eqn (2.1) to the case of molecular liquids has been considered by various workers. We shall adopt the formalism of Powles [**2**] and Appendix 1 shows how this relates to that used by Blum and Narten [**3**]. The molecular structure factor may be written as

$$S_M(Q) = \frac{1}{N_M} \frac{\langle \sum_{ij} \bar{b}_i \bar{b}_j \exp(\mathrm{i}\,\mathbf{Q}\cdot\mathbf{r}_{ij})\rangle}{\left[\sum_i b_i\right]^2}, \tag{2.3}$$

where N_M is the number of molecules in the assembly, $\bar{b}_i$ and $\bar{b}_j$ are the coherent scattering lengths of nuclei i and j at a separation $\mathbf{r}_{ij}$; the summation extends over all pairs i, j (averaged over isotopic composition) of nuclei in the sample and the $\langle\ \rangle$ brackets denote an ensemble average. It is convenient to separate the terms in the series into those arising from correlations in the same molecule (intramolecular terms) and those from correlations in different molecules (intermolecular terms) such that

$$S_M(Q) = f_1(Q) + D_M(Q), \tag{2.4}$$

where the molecular form-factor, $f_1(Q)$ is given by

$$f_1(Q) = \frac{1}{\left[\sum^{N} b_i\right]^2} \sum^{N} b_i b_j\, j_0(Qr_{ij}) \exp(-\gamma_{ij}\, Q^2), \tag{2.5}$$

with $j_0(x) = (\sin\ x/x)$ and $\gamma_{ij} = \frac{1}{2}\langle u_{ij}{}^2\rangle$ where $\langle u_{ij}{}^2\rangle$ is the mean-square amplitude of displacement from equilibrium position due to the normal mode vibration of the molecule; the summation extends over all N atoms in the molecular unit. The $f_1(Q)$ function is equivalent to the diffraction pattern that would be observed for individual molecules in the low-density regime and is solely dependent on the molecular conformation and the coherent scattering length of the constituents. For the D_2O molecules shown in figure 2 this becomes

$$[f_1(Q)]^{D_2O} = \frac{1}{(b_O+2b_D)^2}[b_O{}^2+2b_D{}^2+4b_Ob_Dj_0(Qr_{OD})\exp(-\gamma_{OD}Q^2) + 2b_D{}^2j_0(Qr_{DD})\exp(-\gamma_{DD}Q^2)], \quad (2.6)$$

where r_{OD} is the bond length and $r_{DD} = 2\,r_{OD}\sin\theta/2$ for an intramolecular bend angle, θ. At high Q-values (typically greater than 10 Å^{-1}) the structure factor is dominated by the oscillatory nature of the $f_1(Q)$ contribution so that the molecular conformation can be established from the shape of the diffraction pattern in this region.

At lower Q-values the difference term, $D_m(Q)$, contributes to the overall pattern and may be formally expressed as

$$D_M(Q) = \frac{1}{N_M[\Sigma b_n]^2}\left\langle \sum_{i\neq j} \exp(\mathbf{Q}.\mathbf{r}_{c_{ij}}) \sum_{n_i n_j} b_{n_i}b_{n_j}\exp(i\mathbf{Q}\cdot(\mathbf{r}_{c_{n,i}}-\mathbf{r}_{c_{n,j}}))\right\rangle \quad (2.7)$$

where n labels a nucleus within a molecule, i and j refer to different molecules in which the molecular centre is situated; $\mathbf{r}_{c_n}$ is the intramolecular distance of nucleus n from the molecular centre and $\mathbf{r}_{c_{ij}}$ is the inter-molecular distance between centres of molecules denoted by i and j. This function contains the required information on the 'structure' of the liquid as distinct from that of the molecule. The term 'structure' corresponds to the ensemble average for all the molecules in the system at a specific time. Since the liquid sample is composed of a large number of molecules in continual motion it is more convenient to think of this as a long-time average of the various changing configurations around a single molecule. The structure factor (eqn (2.3)) therefore arises from a 'snap-shot' picture of the ensemble with a shutter-speed which is short compared to the characteristic motion of the scattering centres. In the case of X-rays this static approximation is well satisfied, but this is not

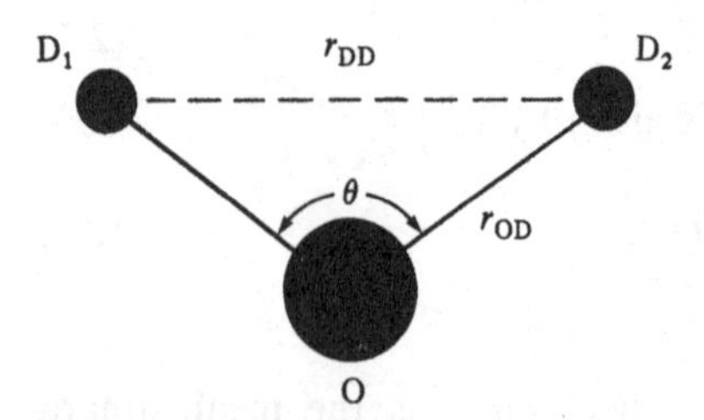

Figure 2. The D_2O molecule.

true for thermal neutron scattering and it is necessary to make appropriate corrections to the observed diffraction pattern; these corrections and the effects that they have on the data treatment are discussed more fully in sections 3, 4.5 and also in Appendix 2.

2.4. *Transformation Relations*

It is, in principle, possible to extract an absolute value of the $S_M(Q)$ function from a diffraction measurement although it is now more usual to carry out the required data analysis procedures directly on the cross-section results as shown in section 4.5. The conversion of $S_M(Q)$ to a real-space representation can be made by a transform analogous to eqn (2.2). The interference part $I_M(Q)$ of the $S_M(Q)$ function is

$$I_M(Q) = S_M(Q) - S_M(\infty) \tag{2.8a}$$

where $S_M(\infty)$ is the asymptotic value of $f_1(Q)$ at high-Q, i.e.

$$S_M(Q) = \frac{\Sigma\, b_i^2}{[\Sigma\, b_i]^2}, \tag{2.8b}$$

so that the pair correlation function $g_M(r)$ may be expressed in the form

$$d_M(r) = 4\pi r\rho[g_M(r) - 1] = \frac{2}{\pi}\int_0^\infty Q\, I_M(Q) \sin Qr\, dQ. \tag{2.9}$$

The $d_M(r)$ and $g_M(r)$ functions represent the total pair correlation function, which is a combination of the separate partial correlation functions and includes sharp peaks due to the intramolecular distances. It is more convenient to remove the intramolecular terms by subtracting the molecular form-factor from the observed data to give $D_M(Q)$, which may be transformed by

$$d_L(r) = 4\pi\rho r[g_L(r) - 1] = \frac{2}{\pi}\int_0^\infty Q\, D_M(Q) \sin Qr\, dQ \tag{2.10}$$

to give the pair correlation functions $d_L(r)$ and $g_L(r)$ for the intermolecular terms only. These composite functions will in general contain contributions from all the partial correlation functions weighted according to the b-values and the relative concentrations, i.e.

$$g_L(r) = \Sigma\, c_\alpha^2 b_\alpha^2 g_{\alpha\alpha}(r) + 2 \sum_{\alpha \neq \beta} c_\alpha c_\beta b_\alpha b_\beta g_{\alpha\beta}(r), \tag{2.11a}$$

which can also be written in more compact form as

$$g_L(\mathbf{r}) = \Sigma\, (2 - \delta_{\alpha\beta}) c_\alpha c_\beta b_\alpha b_\beta g_{\alpha\beta}(r),$$

where the concentration terms are normalized such that

$$\Sigma\, c_\alpha = 1. \tag{2.11b}$$

For neutron scattering by D_2O, the composite function is weighted according to the following proportions:

$$[g_L(r)]^{n/D_2O} = 0.092\, g_{OO}(r) + 0.422\, g_{OD}(r) + 0.486\, g_{DD}(r). \tag{2.12}$$

It is notable that the deuterium atoms make a significant contribution to the total scattering pattern in contrast to X-ray diffraction studies, where it is the oxygen atoms that predominate. It follows that the X-ray studies primarily give information on the correlation of oxygen atoms or the molecular centres whereas the neutron scattering by D_2O (or H_2O) is much more sensitive to the relative alignment of adjacent molecules. It is in this sense that X-ray and neutron experiments give complementary information (see section 6).

The transform equation involves an integral extending to infinity but the available data are limited to a finite range with a maximum value, Q_M. The evaluation of the $d_L(r)$ results is therefore dependent on the value of Q_M and can lead to systematic errors resulting from this truncation. If the results are evaluated directly from eqn (2.10) with the infinite limit replaced by Q_M, this is equivalent to the convolution of the full $I_M(Q)$ or $D_M(Q)$ curve with a step function having a sharp cut-off at Q_M. The resulting $d(r)$ or $d_L(r)$ curve may then exhibit a number of termination ripples and can possibly lead to serious errors in the geometrical interpretation of the data based on the positions of the peaks. The loss of information at high Q-values also leads to a broadening of the main peaks. The real-space resolution which defines the width due to the transform limitations can be shown to be inversely proportional to Q_M. The main principles of the transform procedures have been discussed by Waser and Schomacher [**8**], who show that the termination ripples can be reduced by incorporating a modification function, $M(Q)$, into the transform such that

$$d_L(r) = \frac{2}{\pi} \int_0^{Q_M} Q\, D_M(Q)\, M(Q) \sin Qr\, \mathrm{d}Q. \tag{2.13a}$$

The $M(Q)$ function is chosen to go smoothly to zero at Q_M and a frequently used form is

$$M(Q) = \frac{\sin(\pi Q/Q_M)}{(\pi Q/Q_M)} = j_0\left(\frac{\pi Q}{Q_M}\right). \tag{2.13b}$$

It is clearly beneficial to make Q_M as large as possible, but this imposes conditions on the experimental parameters since the scattering vector for elastic scattering is

$$Q = \frac{4\pi}{\lambda} \sin \theta/2, \tag{2.14}$$

where λ is the incident wavelength and θ is the scattering angle. There is a natural limit, which is dependent on the experimental conditions. The use of short-wavelength neutrons on reactor instruments usually results in a lower beam intensity, so that some compromise is necessary between the Q-range and the statistical accuracy of the measurements. The consideration of both systematic and random errors is fully discussed in Appendix 3, and its relevance to the interpretation of the experimental results is included in the appropriate parts of sections 5 and 6.

2.5. *Difference Techniques*

The structural information obtained from a single measurement does not necessarily give much insight into the properties of the liquid although it can prove useful in checking the validity of computer predictions from molecular dynamics or Monte Carlo simulations. A more detailed study of a single sample can be achieved by variation of external parameters such as temperature and pressure. Although the structural relationships are not usually very dependent on these variables for most liquids, it has been found that liquids with strong association properties, such as hydrogen-bonding, can show significant structural modification. In fact, studies of water appear to show that this liquid is probably more susceptible to structural change than any other liquid with the possible exception of deuterium fluoride [**9**], which exhibits a number of interesting characteristics due to the extreme nature of the hydrogen-bonding interactions. It is therefore convenient to use a formalism based on first-order difference methods to study these changes.

The principles can be illustrated most clearly in relation to temperature variation studies. If the structure factor is known or measured for a reference temperature, T_0, then the first-order difference measurement involves the term,

$$\Delta S_{\mathrm{M}}(Q,T) = S_{\mathrm{M}}(Q,T) - S_{\mathrm{M}}(Q,T_0) \tag{2.15a}$$

and if it is assumed that the molecular conformation is unchanged, this may be written more simply as

$$\Delta S_{\mathrm{M}}(Q,T) = \Delta D_{\mathrm{M}}(Q,T), \tag{2.15b}$$

since the $f_1(Q)$ term of eqn (24) remains constant. The transform relation for the difference function can now be written as

$$\Delta d_{\mathrm{L}}(r,T) = \frac{2}{\pi}\int_0^{\infty} Q\,\Delta D_{\mathrm{M}}(Q,T)\sin Qr\,\mathrm{d}Q \tag{2.16}$$

where $\Delta d_{\mathrm{L}}(r,T)$ represents the change in the real space correlation functions with respect to temperature, and may be written to good approximation as

$$\begin{aligned}\Delta d_{\mathrm{L}}(r,T) &= 4\pi r[\rho\,\Delta g(r) - \Delta\rho(g(r)-1)]\\ &= 4\pi r\rho[\Delta g(r) - \alpha_{\mathrm{v}}(g(r)-1)\,\Delta T],\end{aligned} \tag{2.17}$$

where α_{v} is the coefficient of volume expansion. In the case of strongly hydrogen-bonded liquids the second term in this expression is usually much smaller than the first, so that the transform of $\Delta D_{\mathrm{M}}(Q,T)$ reveals the change in the real-space correlations as represented by $\Delta g(r)$. Corrections for the density term can be applied if necessary, but some workers [**10, 11**] have used the fact that water has an anomalous behaviour with a density maximum so that it is possible to choose two temperatures with an identical density. In this case the results give the isochoric temperature derivative $[\partial S(Q)/\partial T]_\rho$, which shows the effect of temperature for equivalent values of density.

The advantage of the formalism presented here is that the changes in structure can be evaluated without the need for a precise knowledge of the various experimental and analytic correction factors that need to be applied to extract the required information from the observations, since these will be identical for both measurements and the final results will be much less sensitive to the data treatment. The main requirement from the experiment is a high count-rate, to enable the small differences in the diffraction pattern to be measured with sufficient statistical accuracy. A similar formalism can be developed for pressure studies, and both types of experiment have provided valuable information (sections 4.1 and 4.2).

2.6. *Separation of Partial Correlation Functions*

The diffraction data obtained from any sample will normally consist of a composite mixture of individual pair correlation functions. It is clearly desirable to obtain independent information for each of the partial terms if this is accessible to experimental study. In the case of water with two components, the three partial functions can be separated if three independent diffraction measurements are made in which the relative proportions of the different terms are varied. This procedure can be adopted in neutron studies if one or more of the components has two isotopes with different b-values. The principle of isotopic substitution has proved of immense value in the case of neutron measurements on aqueous solutions [**12**] and has provided detailed structural information which would have been almost impossible to obtain by any other means. In the case of water the conditions are not so straightforward, as illustrated in table 1, which lists the relevant neutron parameters for the stable or long-lived isotopes of hydrogen and oxygen. It is unfortunate that the b-values for the isotopes of oxygen are almost identical and therefore offer no possibility for isotopic substitution experiments because the differences are too small to provide a practicable measurement. In contrast, there is a large difference in the b-values of hydrogen and deuterium but there is also a very large incoherent contribution in the case of hydrogen which can give no information on the structural characteristics. This property is particularly useful for inelastic and quasi-elastic studies of proton dynamics but is a major problem for diffraction studies. Despite the obvious difficulties involved in the use of hydrogenous materials, it has been recognized that this is probably the only reliable means of obtaining the required information on the separate partial $g(r)$ functions, and several groups

Table 1. *Neutron scattering data for H and O isotopes*

	b_{coh} [fm]	σ_{coh} [barns]	σ_{inc} [barns]
H	−3.74	1.8	80
D (^{2}H)	+6.67	5.6	2.0
T (^{3}H)	+4.7	2.8	—
O^{nat}	+5.8	4.2	~ 0
^{17}O	+5.78	4.2	—
^{18}O	+6.0	4.2	—

Weighting factors for isotopically pure liquids.
H_2O: $g(r) = 0.193\, g_{OO}(r) - 0.492\, g_{OH}(r) + 0.317\, g_{HH}(r)$.
D_2O: $g(r) = 0.092\, g_{OO}(r) + 0.422\, g_{OD}(r) + 0.486\, g_{DD}(r)$.
T_2O: $g(r) = 0.146\, g_{OO}(r) + 0.472\, g_{OT}(r) + 0.382\, g_{TT}(r)$.

have therefore used the H/D isotopic substitution method. The discussion of these measurements and the associated problems of interpreting the results is considered in section 6.

2.7. *Inelasticity Corrections*

The evaluation of the structure factor from the differential scattering cross-section requires the application of correction terms to allow for inelasticity. The formal representation of $S_M(Q)$ given in eqn (2.3) assumes that the scattering is elastic and no energy exchange between the neutron and the assembly of scattering centres occurs. The static approximation is well satisfied in X-ray diffraction studies but in the case of neutron scattering some account must be taken of the dynamics of the scatterers. A full description of the assembly is given by the dynamic structure factor $S(Q, \omega)$ introduced by van Hove [13] and the transform $G(\mathbf{r}, t)$, which represents the time evolution of the pair correlation function; ω is the frequency corresponding to an energy $\hbar\omega$. Using this notation the structure factor $S(Q)$ is simply the integration of $S(Q, \omega)$ over all ω at constant Q, i.e.

$$S(Q) = \int_{-\infty}^{\infty} S(Q, \omega)\, d\omega. \tag{2.17}$$

The use of a detector at a set position in the reactor experiment (section 3) results in an integration of neutrons entering the detector at a constant scattering angle. If the neutron exchanges energy with the scatterer, the true Q-value will be different from that given by eqn (2.14). Furthermore, the detection efficiency will generally vary for different neutron energies, so that the observed cross-section will be of the form

$$\left[\frac{d\sigma}{d\Omega}(\theta, \lambda)\right]_{obs} = \int_{-\infty \atop [\text{const } \theta]}^{\omega_m} F(k') \left|\frac{k'}{k}\right| S'(Q, \omega)\, d\omega \tag{2.18}$$

where k and k' are the incident and final wave vectors of the neutron, $F(k)$ is the detector efficiency and ω_m is a frequency cut-off point because the incident neutron cannot impart more energy to the system than it already possesses; the expression $S'(Q,\omega)$ is the neutron-weighted dynamic structure factor given by

$$S'(Q,\omega) = \frac{1}{2\pi}\int_{-\infty}^{\infty} dt \exp(i\omega t)\langle\sum_{ij} b_i b_j \exp(i\mathbf{Q}\cdot\mathbf{r}_i(t))\exp(i\mathbf{Q}\cdot\mathbf{r}_j(t))\rangle \qquad (2.19)$$

which incorporates the time-dependent pair correlation functions. It is immediately apparent that the observed values for the cross-section are dependent on the dynamics of the scatterers, and some assumptions must be made in order to estimate the deviations from the static approximation. The first person to tackle this problem was Placzek **[14]**, who used a power series expansion in ω for the treatment of scattering by monatomic fluids. Under these conditions he showed that the cross-section per atom could be written in the form

$$\left[\frac{d\sigma}{d\Omega}(\theta,\lambda)\right]_{\text{self}} = b^2\left\{1 - K\left(\frac{m_n}{M}\right)\sin^2\frac{\theta}{2} - \frac{1}{3}\frac{K_{av}}{E_0} + \dots\right\}, \qquad (2.20)$$

where m_n is the mass of the neutron, M is the mass of the atom, K is a detector constant, E_0 is the incident neutron energy and K_{av} represents the average kinetic energy of the atoms. The term in $\sin^2\theta/2$ is proportional to Q^2, and this is the origin of the frequently quoted 'Q^2 fall-off' in the cross-section for self-scattering. The magnitude of the effect is dependent on the atomic mass of the scatterer and becomes more pronounced for low values of M.

The generalization of this procedure to molecular liquids is complex and there is still some disagreement on the most appropriate method. The difficulty arises from the characterization of the dynamics, because each scatterer is now a part of a larger molecular unit and its motion is not described in terms of simple translational diffusion as in the case of monatomic liquids. The effects of molecular vibration and rotation will influence the scattering characteristics, and although these features will be formally included in the $S(Q,\omega)$ representation, it means that some approximations must be made in order to satisfactorily evaluate the integrals over a much wider range of Q and ω-values. Several groups have worked on this problem and a presentation of the more technical aspects of this development is given in Appendix 2. The main effect is on the self-scattering terms. At low Q-values the cross-section behaves in a similar way to that given by Placzek in which an effective mass M_{eff} characterizes the magnitude of the Q^2 fall-off. For heavy atoms this formalism is adequate for the description of the complete angular range, but the use of deuterium or hydrogen poses major problems since the predicted fall in overall intensity becomes too large in the high-angle (high-Q) region. The importance of the corrections can be most readily appreciated by noting that the cross-section for scattering of a neutron by a free proton must be zero in the backward hemisphere of the laboratory frame unless nuclear charge exchange processes are introduced! In terms of the scattering properties, the light atoms in a molecular unit are therefore neither bound nor free and the temperature-dependent normal mode vibrations of the molecule will therefore influence the scattering process. Fortunately the overall effects on the calculation of the differential cross-section for the self-terms produce a monotonic change, and the associated mass parameters can usually be fitted adequately to the experimental data without causing any distortion of the oscillatory form of the diffraction pattern. It is interesting to note in this context that no attempt is made to evaluate the self-term characteristics in the similar analysis procedure for electron diffraction data, as it is recognized that the inelasticity effects are too large to be reliably evaluated and therefore a smooth 'background' is used to represent this contribution to the overall pattern. The latest calculations of the Placzek corrections for deuterium-containing materials seem to be more than adequate (section 3), but some problems still remain in the treatment of hydrogen (section 6).

A more important factor arises in the representation of the interference terms which give the required structural information. Within the molecular unit there are significant correlations in the motion of the different atoms, and this has a direct effect on the oscillatory structure of the intramolecular contributions to the diffraction pattern. The main effect may be described as an apparent recoil of the whole molecule, although the actual expressions depend individually on the separate intramolecular terms in the $f_1(Q)$ form-factor. The result of this phenomenon is to produce an apparent shift in the Q-scale of the diffraction pattern such that oscillatory shape in the high Q-value region is displaced relative to that corresponding to the static case. If the uncorrected observations are transformed according to eqn (2.9), the peak position will appear at an r-value which is smaller than the correct one. This apparent reduction in the bond length will also occur if a parameter fit is made to the high-Q range using the standard $f_1(Q)$ formulation, and will therefore lead to systematic errors in the conformation of the molecule. It is now normal procedure to analyse the cross-section data by incorporating the corrections to the intramolecular terms in the overall description of the scattering cross-section for each molecule. Although these procedures require extensive computation and careful treatment in relation to the observed cross-section distribution there are various consistency checks (see Appendix 3) which provides safeguards against erroneous deductions, and the effective mass parameters also convey some information about the influence of the local environment on the dynamic processes.

It can be see that slow neutron diffraction studies of deuterium and hydrogen-containing materials are inextricably bound up with the dynamical properties of the assembly. It is the basic aim of the two-axis diffraction study to obtain structural information, and the inelasticity effects are therefore a hindrance to the achievement of this information. It is also clear that the currently established procedures are sufficiently well understood to enable satisfactory information to be extracted from the observations provided due care is taken in the treatment of the data. The magnitude of the corrections for any chosen Q-value decreases as the neutron energy is increased, so this provides another reason for using short-wavelength neutrons where possible. The very short wavelengths available on pulsed source instruments (section 3) suggest that these facilities will offer considerable advantages for future studies but the nature of the corrections is quite different in this work (see Appendix 3), so that more exploratory measurements will be needed before a satisfactory procedure is defined. The development of techniques for handling these corrections will, of course, have benefits for studies of materials other than water.

3. Experimental Techniques

The first neutron scattering experiments on water and ice were conducted by Schull in 1946. A fascinating account of these pioneering studies was reported at an anniversary meeting in 1971 [**15**] indicating the first realization of the important differences in coherent and incoherent scattering between hydrogen and deuterium. Since these early observations, there has been a steadily increasing development of experimental techniques for the investigation of neutron scattering by water. Much of this work has naturally been devoted to inelastic studies and the establishment of the scattering law $S(Q, \omega)$ to define the dynamics of the interaction in order to evaluate neutron transport properties in water moderators. [**16**] The interest in neutron diffraction as a tool for structural studies of liquid water was developed much later. A review by Page [**17**] in the first volume of *Water: A Comprehensive Treatise* [**1**] gives a general survey of neutron scattering by liquid water in the early stages which incorporates diffraction, quasi-elastic and inelastic studies. Most of the available structural information on water at that time had resulted from the extensive X-ray studies of Narten and Levy [**18**], which are reviewed by Narten [**19**] in the same volume. The intervening period has seen a significant improvement in experimental facilities and also the theoretical concepts required for the evaluation of the diffraction results so that the picture has undergone a considerable change.

3.1. *Conventional Diffraction Methods*

The most usual method of investigation is based on the use of a two-axis diffractometer, illustrated in figure 3(a). Neutrons from the reactor core pass through a channel in the shielding wall and are incident on a monochromator crystal which is used to select a chosen wavelength and produces a diffracted beam at the appropriate take-off angle, Θ_M. The monochromatic neutron

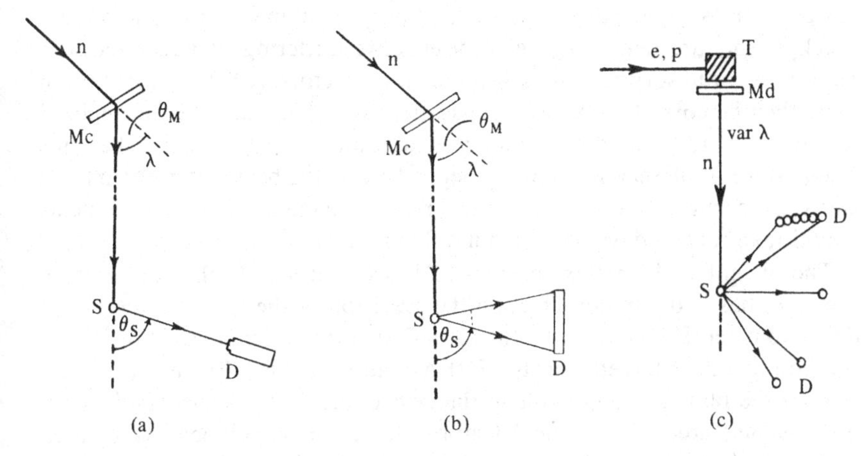

Figure 3. Experimental arrangement for diffraction studies: (a) reactor and detector array; (b) reactor and multidetector; (c) time-of-flight spectrometer with pulsed beam. Mc, monochromator; S, sample; D, detector; T, target; Md, moderator.

beam passes through a transmission monitor counter and is collimated before it reaches the sample position. The scattered neutrons are detected in one or several neutron counters which can be moved over a range of different scattering angles. The intensity pattern is built up by a scan over the accessible angular range using sequential incremental steps of the detector(s) for a set value of the monitor count. Various calibration checks must be made to define the absolute values of the measurements. The incident wavelength is usually determined by placing a powdered sample of a crystalline material with a well-known *d*-spacing at the same position as the liquid sample; the determination of the angular position of three or more Bragg peaks then serves to define both the wavelength and the zero-angle of the diffractometer. The scattering vector for elastic scattering is simply evaluated at each detector setting from eqn (2.14). The count-rate is put on an absolute scale by using a standard scatterer in the form of a vanadium bar. The scattering from vanadium is almost completely incoherent [**5**, **6**] so that the measured count-rate provides a convenient means of defining an absolute scale. Since the relative volumes of the sample and the vanadium are known, the number of scattering centres in the beam can be evaluated and hence the measured diffraction data can be converted to an absolute scale, measured in barns/steradian/molecule using the known value for the total incoherent scattering cross-section of vanadium at that wavelength. Several 'experimental' corrections must be made to the observations in order to determine the total differential scattering cross-section distribution. Since liquids must be placed in suitable containers, separate measurements are required for the empty container. As water is a non-corrosive material it is usual to use a thin-walled container made of either vanadium or aluminium. The raw intensity distributions for 'sample plus container', 'empty container', 'vanadium' and 'background' are combined to give the effective scattering by the sample after allowance for absorption and self-attenuation factors. [**20**] This distribution must then be corrected for multiple-scattering [**21**] and suitably normalized to give the absolute differential cross-section [$\mathrm{d}\sigma\ (\theta,\lambda)/\mathrm{d}\Omega$]. The time required for a full scan is primarily dependent on the beam intensity and the statistical accuracy required. A typical run involving all four measurements would take several days on a medium flux reactor.

The advent of high flux reactors [**22**] designed specifically for research purposes has had a major effect on the precision of the measurements. The diffractometer D4 at the Institut Laue-Langevin, Grenoble, France, and the liquids diffractometer on the HIFA reactor at Oak Ridge, Tennessee, USA, have played a major role in the development of modern facilities for liquid-state studies. The single detector of the earlier experiments is replaced by a multi-detector as shown in figure 3(b). Under these conditions the detector effectively measures a range of scattering angles simultaneously so that the information content obtained from a single setting is much increased. Another development was the use of a hot source [**23**] in the reactor core which shifts the neutron spectrum to higher energies. As a result of this, the beam intensity remains relatively high for the wavelength range below 1 Å. The use of low-wavelength neutrons is advantageous for several reasons. The main one is immediately apparent from eqn (2.14) where it is clear that the available Q-range is progressively extended as λ is reduced so that more information is obtained for the $S_{\mathrm{M}}(Q)$ function. This feature is of particular importance for studies of water, as will be shown later (section 4.1), and the use of variable wavelength methods has some relevance to the inelasticity corrections considered in the next section. These technical developments are continuing, and the new D4B at ILL diffractometer carries this a step further with the installation of a focusing monochromator and two 12° multi-detectors. [**22**] The ultimate stage is represented by the use of a single 140° multi-detector [**24**], as adopted at CEN, Saclay, on the new medium-flux Orphée reactor. The ability to measure the structure factor for a liquid sample to high precision is no longer limited by the available facilities and a run of only 5 h is sufficient to give a good statistical accuracy over a wide Q-range. It is the systematic errors in the treatment of the data which are now the main cause of uncertainties in the final results rather than the random errors arising from the counting statistics. The influence of these factors on the individual measurements will be described in the relevant paragraphs of section 4, but a description of the effects and their origin is incorporated into the general consideration of random and systematic errors presented in Appendix 3.

3.2. *Pulsed Neutron Diffraction Methods*

The production of thermal neutron beams for research purposes is not restricted to the fission process in a nuclear reactor. During the last decade there has been an increasing interest in neutrons produced by accelerator sources. The principles have been reviewed by various authors [**25**] and the development of a wide range of different neutron instruments discussed by Windsor [**26**]. The first use of a pulsed neutron beam for structural studies of the condensed state was based on the Harwell Electron Linac using the total scattering spectrometer (TSS). The principles of the spectrometer are shown in figure 3(c). An accelerated electron beam of about 100 MeV is incident on a heavy metal target and produces a high-intensity pulse of gamma rays which interact with the surrounding material to give fast neutrons by (γn) and (γf) reactions in the target assembly. The burst of neutrons is moderated in a polythene slab so that epithermal and slow neutrons are created. They travel a distance of several metres to the sample position and are scattered through a short flight-path into a number of fixed-angle detectors. The time of flight is recorded for each detector count so that the effective neutron energy and wavelength are determined. Since the electron beam is pulsed at the repetition rate of the accelerator a time-of-flight measurement with a fixed detector determines the diffraction pattern. In this case a range of wavelengths (variable λ) is used for a fixed counter position (constant θ), so the Q-value is similarly defined by eqn (2.14) and the method is complementary to the more conventional experiments using reactor sources (constant λ, variable θ). The initial experiments have been described by Sinclair *et al.* [**27**] and a critical comparison of the reactor and pulsed neutron methods was given later by Dore and Clarke. [**28**] This system has been used by other groups, notably in Japan and the USA, but has now moved to a second phase in which a proton accelerator is used and the neutrons are produced by the spallation process. In this case the incident projectile carries much higher momentum and effectively knocks neutrons and other particles out of the target nucleus. This scheme has several advantages [**29**] and gives a much higher neutron yield for each incident particle. There are already several machines operating on these principles [**30**] but, unlike the reactor system which has been fully optimized, there are further developments which can yield even higher neutron fluxes. [**31**] The most recently constructed instrument of this type dedicated completely to condensed matter research is the Spallation Neutron Source (SNS) [**29**] at the Rutherford–Appleton Laboratory, near Oxford, which first produced neutrons in a test run during December 1984.

The basic experimental measurements on the 'sample plus container', 'empty container', 'vanadium standard' and 'background' are made in a similar manner to the reactor experiments, but different computational techniques must be used to extract the differential scattering cross-section through the evaluation of absorption, self-attenuation and multiple scattering correction factors. The details of these procedures extend beyond the scope of this review but may be found in current literature. [**28**] Additional correction features occur in the time-of-flight method [**32**] since the shape of the incident neutron spectrum must be obtained from the measurements on the vanadium standard and the effective zero of the initial neutron pulse is found to be dependent on the wavelength. [**33**] A further consequence of the measurement is that the corrections for inelasticity [**34**] in the scattering process are totally different from those occurring in the reactor experiments [**35, 36**]; these complications will be considered in section 4.3.

3.3. *Additional Techniques Relevant to Diffraction Studies*

The previous sections have emphasized the value of neutron scattering in obtaining structural information but have also shown that the interpretation of the observations may be complicated by various effects such as incoherent scattering contributions (section 2.1) and the inelastic scattering effects (section 4.5). Although the necessary correction procedures can often be computed to satisfactory level, it is obviously preferable if some experimental verification can be obtained. In this section, some of the peripheral experiments are described which can be used to supplement the information used in the evaluation of the diffraction results.

The two-axis diffraction experiment (figures 3(a) and (b)) involves a total scattering measurement of the neutron flux entering the detector, and further information can be obtained from an energy analysis of the scattered neutron flux. Various neutron instruments are used for inelastic and quasi-elastic studies [**22**] and extensive work has already been conducted with water samples [**37**] However, it is usual to employ longer-wavelength neutrons (typically, $\lambda > 2$ Å) in order to improve the energy resolution of the instrument, so that most of this work is of little direct relevance to diffraction studies using short-wavelength neutrons. In order to make an appropriate test of the routines used to correct the observed diffraction pattern for inelasticity effects, it is necessary to make an approximate measurement of the dynamic structure factor $S(Q, \omega)$ over a wide range of Q and ω values. The triple-axis spectrometer is not suitable for this purpose since it is very inefficient in making a global survey of a wide region of Q–ω space. The most suitable instrument is the rotating-crystal spectrometer, although this is not ideal for operation at very short wavelengths. Some results [**38**] have been obtained at 1 Å using the IN4 instrument at ILL and will be described later (section 4.5). Since the structure factor is simply the integral over all energy transfers, i.e.

$$S(Q) = \int_{-\infty}^{\infty} S(Q, \omega)\, \mathrm{d}\omega \tag{2.17}$$

it might seem that the ideal way of solving the problems in the data treatment would be to determine $S(Q, \omega)$ over an adequate range of Q and ω space. This procedure is possible with monatomic liquids such as argon [**39**] but proves to be totally impracticable for molecular liquids because the vibrational properties of the molecule involve high values of ω and the strong intra-molecular correlations lead to significant oscillatory structure at high Q-values. Even including the advances currently being made with pulsed source instrumentation, it is unlikely that sufficient progress will be made in this area to make these types of investigation feasible, and therefore there will always be a need to use computational techniques to evaluate the inelasticity corrections; the supplementary information obtained will serve primarily to define the acceptable range for the parameter values used in the theoretical formalism. This fundamental limitation is particularly restrictive for samples containing hydrogen but is less serious for deuterium compounds.

The separation of the coherent and incoherent scattering contributions to the measured differential cross-section data, or alternatively, the self and distinct components of the structure factor, also raises some problems. In this case, an experimental procedure is again feasible, based on a spin-analysis of the scattered neutrons. For a sample in which there is no preferential alignment of nuclear spins, the scattered beam can be divided into two parts comprising spin–flip and no-spin–flip components. For an unpolarized beam, a statistical treatment [**40**] shows that the two cross-sections may be written as

$$\left[\frac{d\sigma}{d\Omega}\right]_{nsf} = \left[\frac{d\sigma}{d\Omega}\right]_{coh} + \frac{1}{3}\left[\frac{d\sigma}{d\Omega}\right]_{si} \tag{3.1a}$$

$$\left[\frac{d\sigma}{d\Omega}\right]_{sf} = \frac{2}{3}\left[\frac{d\sigma}{d\Omega}\right]_{si}. \tag{3.1b}$$

The subscripts refer to spin–flip (sf), no-spin–flip (nsf) and spin incoherent (si) cross-sections; isotopic incoherence does not produce spin–flip processes so the method only separates the spin-incoherent part. The separate cross-sections may be determined with a polarization triple-axis instrument using techniques originally developed by Moon, Riste and Koehler. [**41**] The diffractometer D5 at ILL [**22**] is based on similar principles and uses a polarized incident neutron beam produced by a polarizing monochromator crystal and a similar analyser crystal to determine the polarization state of the scattered neutrons as shown in figure 4. Experimental results for liquid D_2O have been reported by Dore, Clarke and Wenzel, [**42**] which showed that a satisfactory separation could be achieved but also revealed some difficulties due to depolarization effects and energy resolution problems in the third scattering. The results were interesting, but suggested that much greater statistical accuracy would be required if the method were to be of use of hydrogenous materials. This technique could be highly advantageous in

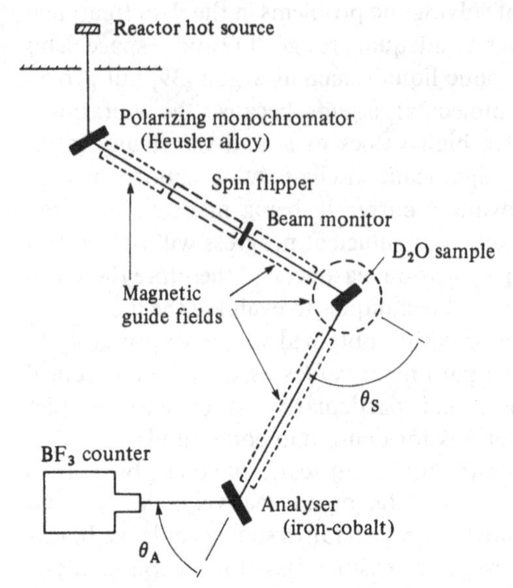

Figure 4. The polarization triple axis diffractometer D5 used for studies of liquid D_2O.

studies of H/D isotopic substitution, since the large contribution due to incoherent scattering by the hydrogen could be substantially reduced. It is possible that further technical advances in the use of white-beam polarization instrumentation **[43]** on pulsed source facilities could re-open this field of investigation, although count-rate limitations could continue to be a major difficulty since the sample scattering must be kept low to avoid a large contribution from multiple scattering. The results of polarization measurements are particularly sensitive to multiple scattering because two spin–flip scattering events will be detected as a no-spin–flip component. The relevance of this type of investigation can be judged more appropriately after reading section 6.2, which describes the present status of work on the partial $g_{\alpha\beta}(r)$ correlation functions using isotopic substitution techniques.

4. Studies of Liquid D_2O

It is natural that most experimental work has been conducted on liquid D_2O due to the large contribution from the incoherent scattering which arises in the case of H_2O. Several different groups have made measurements of neutron diffraction from liquid D_2O at room temperature. The statistical precision achieved in these experiments has increased significantly through the development of improved instrumentation and there has been a corresponding development in the methods of analysing the data; the essential elements of the new results are presented in section 4.1. The improved beam intensity has also meant that new opportunities for difference measurements have been created. The most important series of experiments has been made to investigate temperature-dependent effects in both the normal and under-cooled liquid phases; these are described in section 4.2. In contrast with other non-hydrogen-bonded liquids, there is a significant change in the structure factor with variation of temperature, and this can be interpreted in terms of the structural rearrangement of neighbouring molecules. Less work has been carried out on pressure variation but the available data (section 4.3) give useful information on the variation in hydrogen-bond conformation due to density changes. Another closely related experiment concerns the study of vapour-deposited amorphous D_2O-ice. In this case the molecular motion is drastically reduced and the spatial correlations can be more closely identified; the formation of a random hydrogen-bonded network is shown to be related to the disordered structure in the liquid.

The most interesting development in recent years has been the attempt to obtain more complete information on the separate partial pair correlation functions $g_{HH}(r)$, $g_{OH}(r)$ and $g_{OO}(r)$ by the precise study of scattering from H_2O/D_2O mixtures. These are difficult experiments due to the dominating contribution of incoherent scattering from the hydrogen atoms, and the data interpretation is further complicated by the enhanced inelasticity corrections. The three groups working in this field have reached varying conclusions and it is necessary to present a critical examination of the reasons for these apparently differing views. A comparison of the experimental techniques and analysis procedures is not straightforward but it is possible that a relatively simple resolution of the discrepancies may be within sight. This controversial area remains one of the most important aspects of current work on water, but there are some pointers from other studies which may also help to clarify the present anomalies.

Since the earlier review [**17**] published in 1972 a number of papers have dealt with the measurement and analysis of neutron diffraction results for water; the compilation of the new experimental work on liquid D_2O is briefly summarized in table 2 with some of the characteristics of each particular investigation.

4.1. *Studies of D_2O at Room Temperature*

The review by Page [**17**] sets out the basic formalism for analysis of the early reactor experiments in which particular emphasis was placed on the need to obtain a wide Q-range to give the required information. This sometimes necessitated the combination of data sets taken with different incident wavelengths, and the problems resulting from the inelasticity effects had not been fully realized at that time. A detailed account of the Harwell results was given by Page and Powles [**44**] and several model descriptions based on different schemes for describing the local orientational correlation were compared with the data. None of the chosen models was found to give

Table 2. *Neutron diffraction experiments on liquid and amorphous* D_2O

Authors	Ref.	Method	λ (Å)	Q-range (Å^{-1})	Sample	Variables
Page & Powles	44	R	{0.85, 1.05}	0.8–10	Liquid	Ambient
Narten	45	R	1.1	0.4–10	Liquid	Ambient (25 °C)
Walford, Clarke & Dore	46, 47	R	{0.35, 0.70}	0.5–30	Liquid	Ambient (20 °C)
Walford & Dore	48	R	{0.84, 1.13}		Liquid	Temperature (11–79 °C)
Gibson & Dore	49	R	{0.5, 0.7, 1.07}	0.5–21	Liquid	Temperature (11–75 °C)
Egelstaff *et al.*	10	R	2.4	1–4.5	Liquid	Temperature (−1.7–23.5 °C)
Ohtomo, Tokiwano & Arakawa	51	L	Var.	1–20	Liquid	Ambient (15°)
Ohtomo, Tokiwano & Arakawa	52	L	Var.	1–20	Liquid	Temperature (25–95 °C)
Bosio *et al.*	53	R	0.70	0.2–10	Under-cooled liquid	Temperature (−14.5–21 °C)
Neilson, Page & Howells	54	R	(1.0)	0.5–5.5	Liquid	Pressure (850 bar)
Wu, Whalley & Dolling	55	R	1.1	0.6–10	Liquid	Pressure {< 9.1 kbar, 25 °C; < 15.6 kbar, 85 °C}
Neilson	56	R	0.70	(0.5–15)	Liquid	Pressure (< 6 kbar)
Wenzel, Lindstrom-Lang & Rice	77	R	0.86	1–12	Amorphous solid	(Vapour deposition)
Chowdhury, Dore & Wenzel	78	R	0.70	0.8–19	Amorphous solid	(Vapour deposition)

satisfactory agreement with the observations. Another early measurement was made by Narten [**45**] and was interpreted in terms of a correlated local neighbours model which permitted the $g_{OD}(r)$ and $g_{DD}(r)$ partial functions to be evaluated on the basis of certain assumptions. These early studies have now been superseded by more extensive measurements.

The importance of the experimental conditions can be illustrated by the results of Walford [**46**] for a series of reactor experiments at different wavelengths; these are shown in figure 5 as a composite set of curves. The best conditions are obtained for short-wavelength neutrons, where a wide Q-range can be covered and the general fall-off in intensity at high Q-values due to inelasticity corrections is minimized. Unfortunately, the statistical accuracy is much lower for this type of measurement due to the lower flux. At longer wavelengths, the flux is higher but the diffraction pattern is spread over a wider angular range, giving better instrumental resolution but also making the higher Q-value information inaccessible. As can be seen from figure 5, the oscillatory structure extends up to 15 Å^{-1} and beyond, so the use of wavelengths much greater than 1 Å leads to a substantial truncation of the available data. A further consequence of the short wavelength is that the general fall-off as a function of Q is reduced and the separation of self and interference terms is more easily achieved. In the early 1970s the inelasticity corrections were applied only to the self-terms using a modified Placzek expansion formalism which effectively leads to a Q^2 fall-off in the high-Q region [**44**] characterized by an effective mass parameter. It was soon

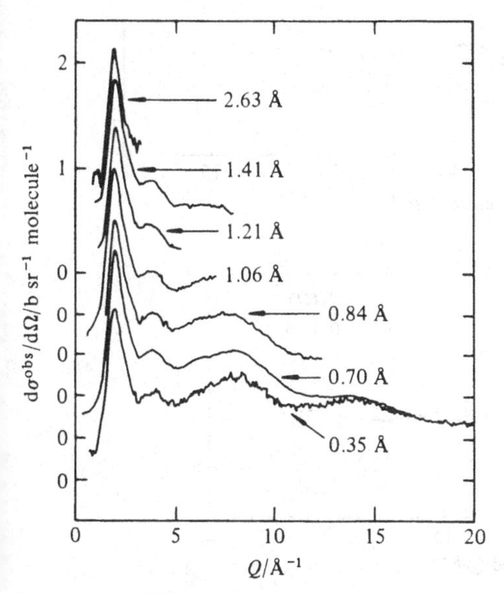

Figure 5. The differential cross-section distribution for liquid D_2O obtained with incident neutrons of various wavelengths [**45**].

realized that this was inadequate for scattering from such a light atom as deuterium and in some cases involving hydrogen could even lead to predictions of a negative cross-section!

The variable wavelength results [**46, 47**] also showed an apparent displacement of the oscillatory structure at high Q-values which was ascribed to molecular recoil. Since this part of the $S_M(Q)$ function is predominantly due to the $f_1(Q)$ form-factor it has a direct bearing on the determination of the molecular bond-lengths r_{OD} and r_{DD} through eqn (2.6). Walford, Clarke and Dore [**47**] used a kinematic correction term to compensate for the recoil effect by introducing a Q-dependent factor into the Q-scale to give agreement between the two data sets. The results are shown in figure 6 and lead to a value of 0.98 Å for r_{OD}. A more detailed study of recoil effects in neutron diffraction measurements was undertaken for the simpler case of nitrogen, studied in both the liquid [**58**] and gas [**59**] phases. The results of these investigations showed that the molecule recoil effect leads to an apparent shortening of the intramolecular bondlengths and could be correctly predicted on the basis of a calculation to determine the effective mass of the scattering centres. It therefore follows that an exact treatment of the dynamical effects is required if precise values for the molecular conformation are to be obtained

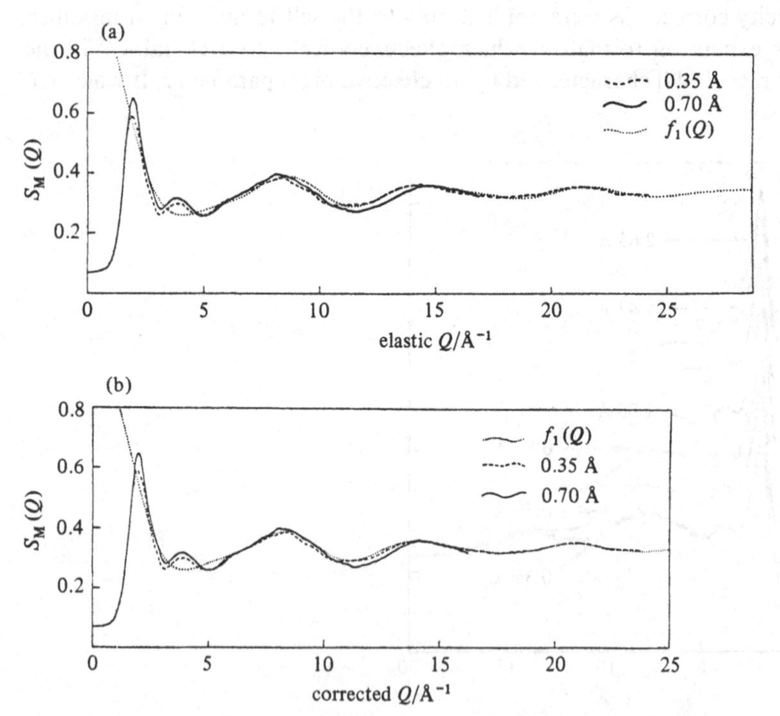

Figure 6. Comparison of the experimental $S_M(Q)$ function for liquid D_2O at 0.35 and 0.7 Å: (a) uncorrected Q-scale; (b) corrected Q-scale.

which are free from systematic error. Several groups tackled this general problem, which has been reviewed in articles by Powles [35] and Egelstaff and Soper. [60] A re-analysis of the UKC data [47] by Powles [60] with a more complete treatment of the recoil effects on the interference terms arising in the $f_1(Q)$ formulation gives a value of 0.983 ± 0.008 Å for the intramolecular bond length, r_{OD} which is substantially in agreement with the value obtained earlier from the kinematic recoil treatment. In addition the r_{DD} distance was found to be 1.55 ± 0.02 Å, giving a mean bend angle, θ_{DOD} of 104 ± 2 degrees. These results are particularly interesting when compared with values from the crystalline, amorphous and vapour phases (see section 5.5). A similar approach was adopted by Blum, Narten and Rovere [36] using a different formalism and was found to give satisfactory agreement with the neutron data for the shorter r_{OD} value from electron diffraction, but no attempt was made to optimize the fit.

It can be seen from the cited examples that a full inelasticity treatment is required and the conversion of the measured differential cross-section, $d\sigma(\Theta)/d\Omega$, to the required structure factor $S_M(Q)$ is not a simple matter. In more recent work on deuterium- (or hydrogen-) containing materials it has been usual to evaluate the corrections for the cross-section computation and to fit the self and intramolecular contribution directly to the observed data in order to determine both the molecular parameters and the variables of the scattering process simultaneously. However, these complex preliminary steps provide no information on the liquid structure and serve only as a first step in the evaluation of $[d\sigma/d\Omega]_{inter}$ and the $D_M(Q)$ function through eqns (2.4) and (2.5). Results for the $D_M(Q)$ function and the corresponding transform, $d_L(r)$ will be presented in later sections, but an example of the early work [46] is shown in figure 7 for the $d(r)$ function obtained from the transform of the

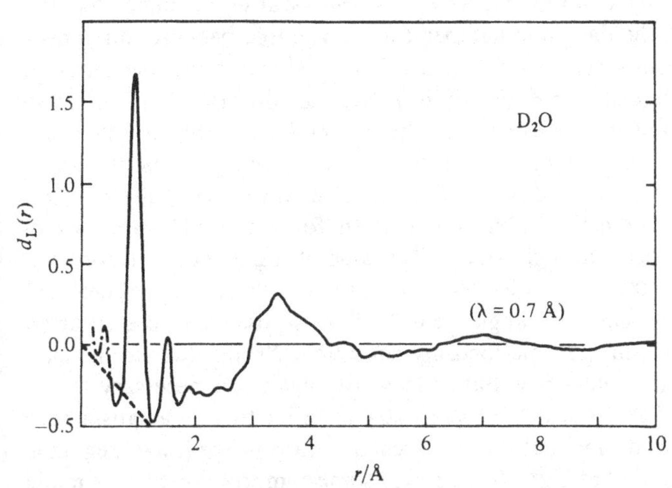

Figure 7. The $d_L(r)$ function obtained from early neutron diffraction measurements of liquid D_2O [45].

full interference pattern. The figure shows two sharp peaks at approximately 1.0 Å and 1.6 Å corresponding to the intramolecular terms, but the rest of the curve is relatively featureless, with an initial broad peak at approximately 3.5 Å possessing a small hump on the low r-value side and followed by a few further oscillations. The results obtained by various groups are in satisfactory agreement concerning these general features, although there are some variations in the magnitudes of the peaks and valleys which probably result from a different treatment of the raw data. It is inevitable that correction methods used for multiple-scattering and for subtraction of the intra-molecular scattering contribution will lead to systematic uncertainties in the derived $D_m(Q)$ and $d_L(r)$ functions. In practice, several consistency checks can be applied at each stage of the data processing. The accuracy of the final $d_L(r)$ results is of importance for the comparison with simulation results, but the combination of statistical and systematic errors makes it difficult to give precise estimates of the magnitudes. Some aspects of the error analysis and the corrections for systematic errors are given in Appendix 3. The interpretation of the $d_L(r)$ data is deferred until section 5.

The advent of pulsed source facilities led to further studies of D_2O at room temperature. Early work by the UKC group [**51**] on the Harwell Linac showed that measurements could be made up to Q-values of 30 Å^{-1} and that the inelasticity corrections for the self terms were much simpler at high Q-values than for the corresponding reactor measurements. However, the full theoretical treatment of inelasticity for this case due to Powles [**34**] emphasized the dependence on other factors such as the shape of the incident spectrum and the dependence on the geometrical ratio of the first and second flight path. The effects of these corrections are clearly seen in some recent calculations by Howells. [**61**] Some attempts were made [**62**] to study these effects explicitly by variation of the counter position for a 90° scattering angle, but the interpretation of the data was hampered by a variable background contribution and the results were never finalized. In principle, it is possible to obtain much more accurate determination of the molecular conformation from these measurements, but the corrections due to molecular recoil introduce additional uncertainty which cannot be properly assessed without further information. Other groups [**51, 52, 63, 64**] have since used pulse neutron methods to determine the diffraction pattern for a lower Q-range using detectors in the forward-angle sector. The time-of-flight spectrum for each detector covers a range of Q-values defined by the wavelengths of the detected neutrons. At large scattering angles (e.g. 150°) it is possible to use an array of several counters in the time-focusing condition [**27**] in order to increase the overall detection efficiency, but at forward angles it is necessary to use Soller slits to define the scattering geometry in order to give a satisfactory Q-resolution. This means that the statistical accuracy in the low-Q region is much poorer than that at high Q-values and some compromise must be made in the design of the instrument. Furthermore the full Q-range can only be obtained by combining data sets from different counters. However, it is clear from the treatment of the inelastic corrections that there is an advantage in making measurements with short-wavelength neutrons and scattering at small angles. Ohtomo, Tokiwano and Arakawa [**51**] have reported results for D_2O at 15 °C covering a Q-range of 1–20 $Å^{-1}$. The structure factor measurements were found to be in good agreement with data [**45**, **46**] taken on reactor instruments over the 1–8 Å Q-range apart from a broadening of the main diffraction peak.

More recently Soper and Silver [**63**] have used a similar method for the study of H_2O/D_2O mixtures (section 6.1) but also include an interesting discussion [**63**] of reactor and pulsed neutron measurements. The comparison of two data sets for each method is given in figure 8, which provides a clear illustration of the differences. The main point concerns the high-Q behaviour, where the overall level remains constant in the pulsed neutron observations and there is no fall-off as for the reactor case. However, it is also important to note that this situation does not apply at lower Q-values, so that the inelasticity corrections and the effects of the spectrum shape are consequently more significant, as shown in figure 9 from Ohtomo *et al.* [**51**] Another feature which is apparent from figure 8 is the reduction of more than 30% in height of the first diffraction peak due to the lower resolution of the forward angle counters. It is difficult to present a critical assessment concerning the relative advantages of the two methods without including technical details which extend beyond the scope of this review. It is clear, however, that both techniques provide valuable means for investigating the basic problems and

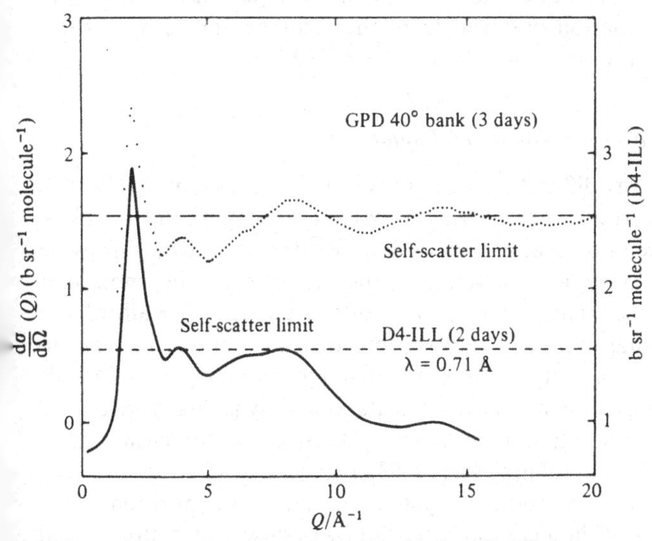

Figure 8. A comparison of neutron scattering from D_2O using reactor and pulsed neutron techniques. [**64**]

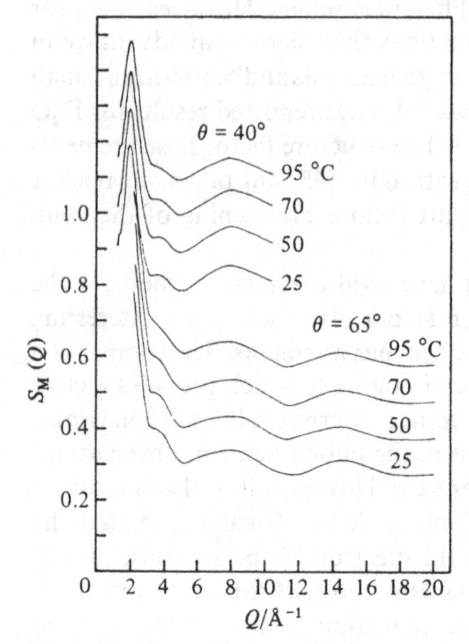

Figure 9. The overall shape of the corrected neutron intensity for pulsed neutron studies of D_2O. [**51**]

should be regarded as complementary methods rather than alternatives. The main advantages of the pulsed neutron technique will be achieved when the second generation of instruments, based on the use of a closely packed scintillator array at low angles such as SANDALS [**65**], is brought into operation. The information obtained from the available data for all these experiments will be discussed in sections 6.1 and 6.2.

4.2. *Temperature Variation Studies of Liquid D_2O*

A comprehensive X-ray diffraction study of water was carried out by Narten and Levy [**18**] and further work presented by Hajdu, Lengyel and Palinkas. [**66**] It was therefore clear that corresponding neutron data would provide valuable complementary information on the changes in the molecular orientational properties. The first neutron results, presented by Walford and Dore [**48**], showed that many of the problems arising from uncertainties in the correction for inelasticity were eliminated by a first-order difference technique (section 3.2). The initial experiments were only partially successful due to wavelength contamination, and more extensive sets of measurements were later reported by several groups. [**49–53**] The most obvious effect of the temperature variation is to produce a systematic shift in the position of the main diffraction peak. The peak position $Q_0(T)$ is shown in figure 10 and exhibits a monotonic behaviour which extends well into the under-cooled

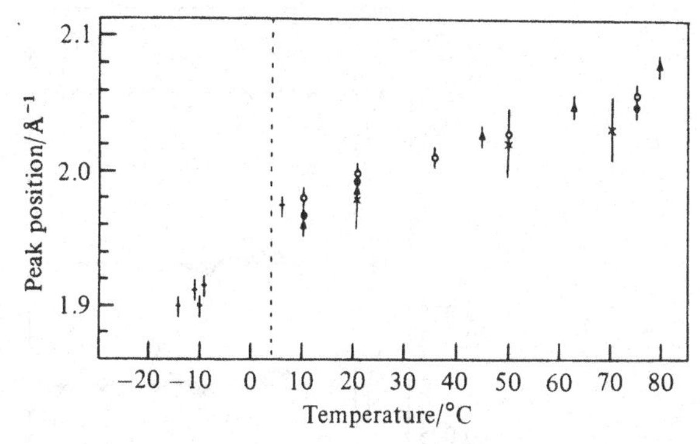

Figure 10. The variation of the position of the main diffraction peak for liquid D_2O as a function of temperature; the broken line represents the melting point of D_2O.

region. The changes are not simply due to density variation, since the decrease in density above 11 °C would be expected to increase the length scale of the intermolecular contributions and lead to a decrease in $Q_0(T)$. The changes are also much greater than for any other molecular liquid that has been studied so far, including some that exhibit strong hydrogen-bonding effects. In this respect, water may be described as having a 'fragile structure' which is sensitive to small changes in external variables such as temperature. The changes in the X-ray patterns **[18]** were also considered, but more recent experimental work by Bosio, Chen and Teixeira **[11]** has improved the precision. The results were taken in both the normal and under-cooled liquid phases and analysed in terms of the isochoric temperature derivative.

Although the shift of the peak position for the neutron measurements on D_2O provides a good 'fingerprint' for the effective temperature of the liquid, it does not directly convey any structural information. In order to extract the required $\Delta D_M(Q, T)$ function a high statistical accuracy is required. A typical set of measurements made on the high-flux diffractometer D2 at ILL is shown in figure 11. Superficially, the cross-section distributions appear very similar, but there are significant differences which are determined by careful subtraction of the two curves after correction for density variation and container expansion. A set of values for the $\Delta D_M(Q, T)$ function is shown in figure 12 and it is pleasing to note that measurements at several different wavelengths are in excellent agreement; the shape of the $D_M(Q)$ function is also shown for comparison. The transform of these curves gives the differential function $\Delta d_L(r, T)$ which is shown in figure 13 and may be compared with the total $d_L(r)$ function shown at the bottom of the diagram. There are four regions of the $\Delta d_L(r, T)$ function which are representative of different characteristics of the structural relationships. At short r-values (< 1.5 Å), where $g(r) \sim 0$, the differential curve is solely dependent on the change in density, $\Delta\rho$, and this provides a convenient check on the calibration of the $\Delta d_L(r, T)$ scale. At

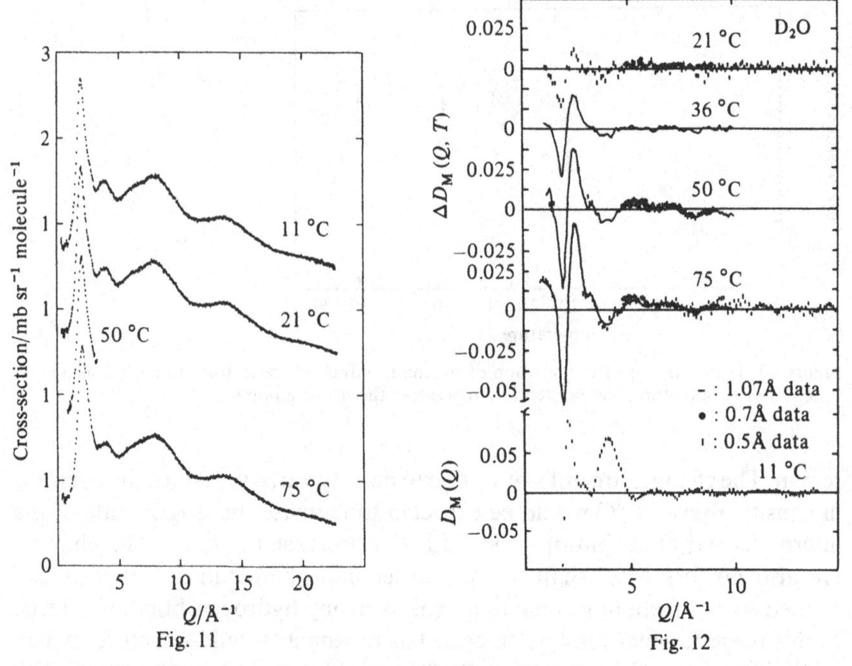

Figure 11. The differential scattering cross-section for liquid D_2O at various temperatures using 0.5 Å neutrons.

Figure 12. The temperature difference function $\Delta D_M(Q, T)$ for liquid D_2O.

low values ($1.5 < r < 2.5$ Å), there is a relatively sharp negative peak followed by an equally sharp positive peak. If it is assumed that the molecular conformation is not influenced by a temperature change of 50 °C, this behaviour can be best explained by a slight increase in the intermolecular hydrogen-bond distance r_{OD} and an increased spread of hydrogen-bond bend angles, as shown schematically in figure 14. The intermediate region ($2.5 < r < 6$ Å) has a further oscillatory structure which requires additional information (section 4.4) for interpretation. In the high r-value region the oscillatory form is out of phase with that of $d_L(r)$. This is partly due to the term in the density variation, $\Delta\rho$ (eqn (2.17)), but also shows a reduction in the long-range order due to the additional disorder created by thermal motion. Later measurements by Bosio *et al.* **[53]** have extended this study to lower temperatures and into the under-cooled liquid phase. The lowest temperature reached was − 14 °C, which corresponds to 20° below the normal freezing point. The results showed a very similar behaviour indicating a systematic change over the whole temperature region even though this includes the point of maximum density.

Similar work has been carried out by other groups. Egelstaff, Chen and collaborators **[10]** have used the isochoric temperature derivative to eliminate

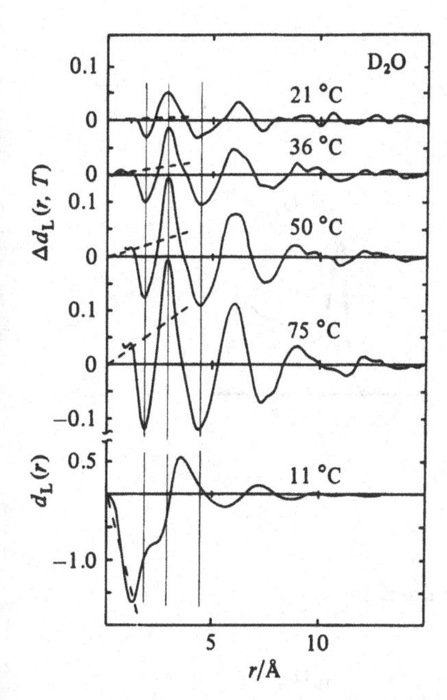

Figure 13. The $\Delta d_L(r,T)$ function obtained from figure 12.

the $\Delta\rho(T)$ term (section 2.5). The data were limited to a Q-range of 4.5 Å^{-1}, which is insufficient to give detailed information in r-space by the transform relation. The data are similar to those obtained from the other studies and were compared with predictions [**66**] for $\partial S(Q)/\partial T$ from the MCY potential. [**67**] the shape of the differential curves was in good agreement with the computer simulation but of much larger magnitude, as shown in figure 15, indicating again the high sensitivity of the structure to external factors.

The pulsed neutron method has been used by Ohtomo *et al.* [**52**] to obtain similar information over the temperature range 25–95 °C, but the adopted method of analysis was different. Using an extension of the formalism for orientational correlation, they developed a way of representing the liquid as a mixture of individual molecules and tetrahedrally coordinated pentamer clusters as shown in figure 16. Since the required form-factors can be expressed completely in terms of Q, the only variable is the relative proportion of each of the two components. It is assumed that the clusters are uncorrelated with each other and the fraction, x_5, of pentamers can then be evaluated from the experimental data. The results show that x_5 decreases monotonically from a value close to unity at 15 °C to approximately 0.55 at 95 °C. The model is instructive because it shows clearly that the subsidiary peak in $S_M(Q)$ at approximately 4 Å^{-1} is particularly sensitive to the x_5 value. This was the region where the models used by Page and Powles [**44**] were unable to give satisfactory agreement with the experimental data available at that time.

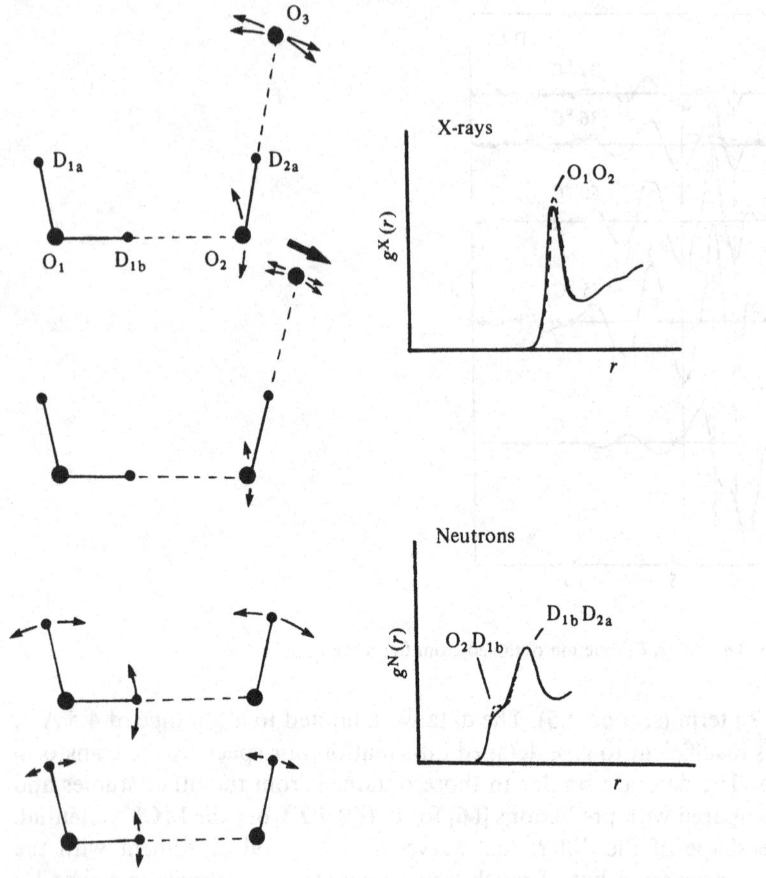

Figure 14. The characteristics of the hydrogen-bonding configuration and expected variation with temperature.

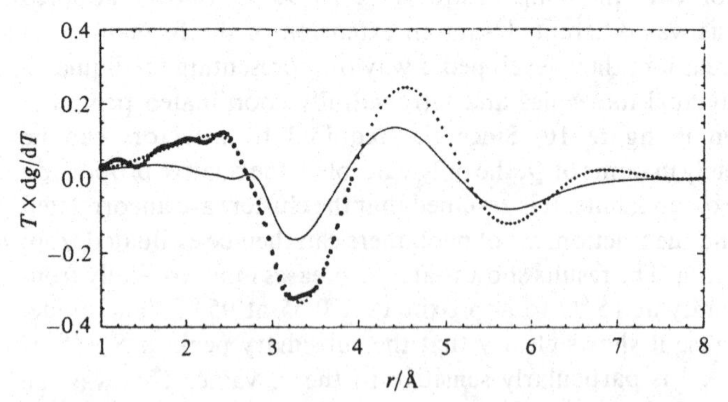

Figure 15. Comparison of the isochoric temperature derivative $(\partial S/\partial T)_\rho$ for D_2O and MCY predictions.

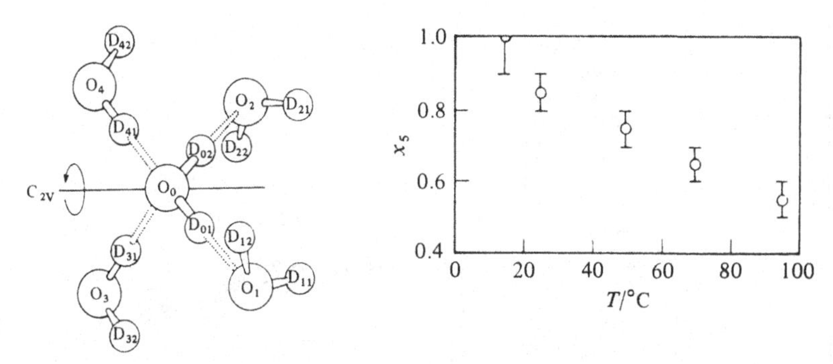

Figure 16. The hydrogen-bonded pentamer unit of Ohtomo *et al.* **[52]** and the variation of the parameter, x_5, with temperature.

Comparison of the position and height of the first diffraction peak was also found to be in good agreement with other results (figure 10), confirming the basic features of the different measurements. It is interesting to note the success of this model description, but the use of a two-component representation obviously has limitations and the actual behaviour of the real fluid would be expected to show some variations from this picture. Further discussion of the geometrical relations is deferred until section 5.3.

4.3. *Pressure Variation Studies*

Experimental studies of liquids under pressure require specially designed sample containers that can withstand the high pressures but which do not contribute a large intensity to the scattered neutrons. These requirements are obviously incompatible, and two different approaches can be made to optimize conditions for the measurements. The method adopted by Neilson, Page and Howells **[54]** was to use a pressure cell made from a titanium–zirconium null matrix material which has a high tensile strength. Since Ti has a negative scattering length, the alloy is composed of the appropriate proportions to make the mean coherent scattering length of the alloy equal to zero. If the two types of atom are randomly distributed in the polycrystalline lattice the scattering should be completely incoherent. In practice it is found that there is some preferential clustering and therefore a small intensity from Bragg peaks is usually present, but this is not of sufficient magnitude to interfere with the observations. Some initial low-Q measurements on liquid D_2O at 1 bar and 850 bar are shown in figure 17. The increase in pressure causes the main diffraction peak to move to higher Q-values. This effect is partly due to the increase in density but is probably enhanced by the effect of pressure on the hydrogen-bond network. Since the tetrahedral nature of the bonds produces an open structure as in crystalline ices, **[68]** the increase in pressure and associated increase in density can only be achieved by

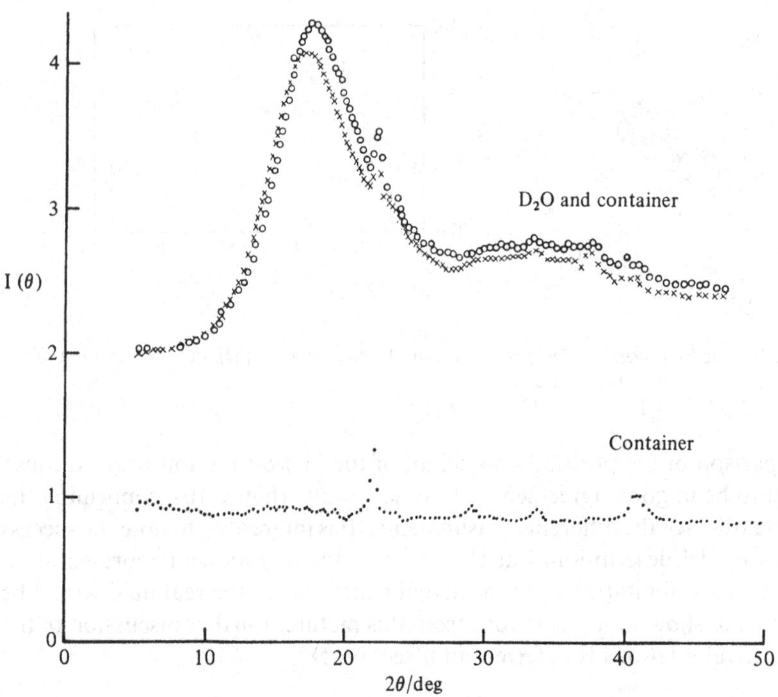

Figure 17. Comparison of neutron scattering from liquid D_2O at 1 and 850 bar. [**54**]

changing the angular relations of the hydrogen-bonds to form a more compact structure. The authors use an expression due to Egelstaff [**69**] which relates the scaling of the intermolecular distances to the slope of the structure factor, i.e.

$$-\rho k\mathrm{T}\left(\frac{\partial S(Q)}{\partial P}\right)_T = \frac{QS(0)}{S}\frac{\partial S(Q)}{\partial Q}.$$

This approximate relation applies to monatomic liquids and is not strictly valid for molecular fluids, but the experimental data showed surprisingly good agreement in both shape and magnitude.

A much wider pressure range was used in the extensive studies of Wu, Whalley and Dolling. [**55**] In this case the scattering from the container walls of an aluminium alloy cell was reduced by a collimated incident beam which was narrower than the sample. Although the main bulk of the container is not in the incident beam, the scattering geometry is more complex and the active volume of liquid varies with the scattering angle. The method is useful for differential measurements but cannot yield accurate absolute values due to the systematic errors arising from the geometrical corrections. This is immediately apparent from the four different runs reported in the paper [**55**], since the final $S_M(Q)$ curves show significant variations in overall shape, but the changes are well defined within each individual run. In the data treatment it was necessary to use other measurements at 1 bar as a suitable reference to normalize the results. This procedure works satisfactorily, so that the systematic changes are observed for pressures up to 15.6 kbar. A more recent set of neutron diffraction measurements has been reported by Neilson **[56]**, who has also given a critical account of the most suitable materials for high-pressure work of this kind. The data were taken at ambient temperature over a pressure range up to 6 kbar using a modified Ti/Zr cell. A particular feature of those results concerns the behaviour of the intramolecular OD peak, which shows a systematic increase in width of approximately 50 per cent from 1 bar to 6 kbar although the peak position remains constant at 0.98 Å. The results of all these experiments are considered in section 5. Corresponding measurements using X-rays have also been reported by Gaballa and Neilson **[57]** on samples of liquid H_2O and D_2O. The relatively small difference in the shape of the double hump in the main diffraction peak is found to increase as the pressure is increased to 2 kbar. The transformed data suggest that the width of the intramolecular peak decreases in the case of light water but increases for heavy water in agreement with the later neutron results. The reason for this behaviour is not understood, but appears to correlate with other observations of water under pressure.

4.4. *Amorphous Ice*

It is clear that the structural properties of water are critically dependent on the properties of the hydrogen bonding. Rice **[70]** suggested that the key to the understanding of these properties could be found in the study of amorphous ice, which is a disordered solid and therefore differs from water primarily in terms of the dynamics of the molecular motion. Amorphous ice is readily formed by vapour deposition and has been known for a considerable time. **[71]** In recent years it has been prepared directly from the liquid phase by Bruggeller and Mayer **[72]** using a cryo-pumping technique and also by Dubochet and Lepault **[73]** using a thin-film system cooled in liquid nitrogen. More recently Mishima, Calvert and Whalley **[74]** have also obtained amorphous ice by a different route using pressure relaxation. For many years it had been thought that a continuity of state between water and disordered ice was not possible **[75]** because amorphous ice converts to a crystalline ice Ih or Ic above 135 K. The new results show that vitrification of water can be achieved if the cooling rate is sufficiently rapid or conditions are controlled by other means. There is some evidence suggesting that the behaviour of the material prepared by different techniques may not be equivalent, and Whalley **[74]** suggests that there may be a continuous gradation covering a range of densities.

Rice and collaborators have made a series of measurements using both diffraction and spectroscopic techniques to study the properties of vapour-deposited amorphous ice. The results have been reviewed by Sceats and Rice **[75]** in a comprehensive survey which concentrates mainly on spectroscopic measurements. X-ray studies by Narten, Venkatesh and Rice **[76]** on a-D_2O provided data on oxygen–oxygen correlations and suggested a high- and low-density form, but subsequent work was unable to reproduce the high-density state, which may have been due to special deposition conditions. **[75]** It seems likely that the most usual form prepared by vapour deposition is of low density (0.94 g cm^{-3}) as studied in the neutron diffraction experiments of Wenzel, Linderstrom-Lang and Rice **[77]** on a-D_2O ice. These results confirmed the expected hydrogen-bonded structure and showed that the molecular correlations were much more prominent than for the liquid phase. Further experiments by Chowdhury, Dore and Wenzel **[78]** improved the statistical accuracy and covered a wider Q-range. The $S_M(Q)$ data are shown in figure 18 with the $f_1(Q)$ fit for the single molecule which has an intramolecular bondlength r_{OD} of 1.00 Å $\pm$0.01, which is a little longer than that of the liquid but shorter than that of ice Ih. The main diffraction peak is much sharper than for liquid D_2O and occurs at the lower Q-value of 1.70 $Å^{-1}$. At higher Q-values the curve has a more marked oscillatory behaviour which extends over a wider range. These features indicate that the spatial properties of a-D_2O are more prominent than for the liquid due to the static nature of the disordered network. This is clearly illustrated by the comparison of the $D_M(Q)$ data for each phase, shown in figure 19, and also the $d(r)$ functions obtained from the transform which are given in figure 20. Several well-defined peaks due to intermolecular distances are observed at low r-values in the amorphous

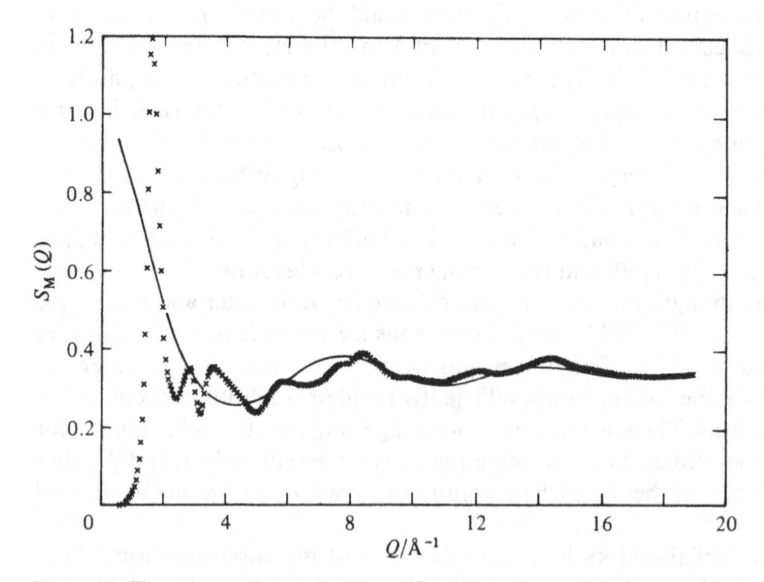

Figure 18. The structure factor, $S_M(Q)$ and fitted form-factor $f_1(Q)$ for amorphous D_2O.

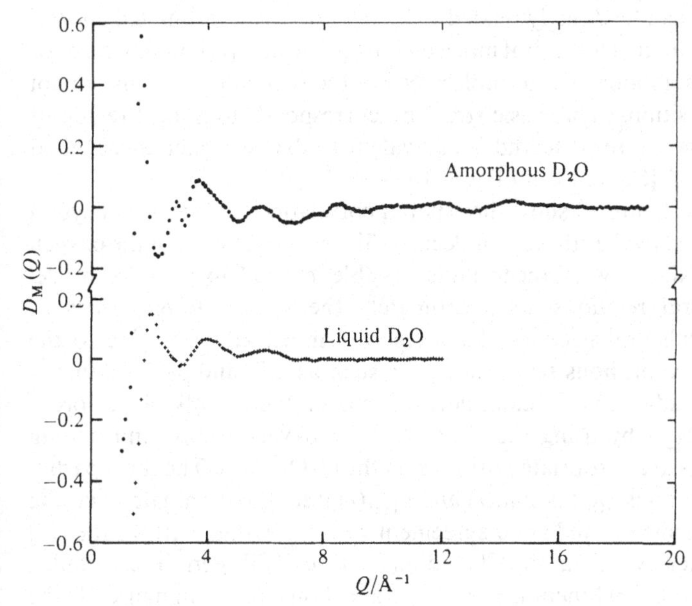

Figure 19. The $D_M(Q)$ function for a-D_2O and liquid D_2O.

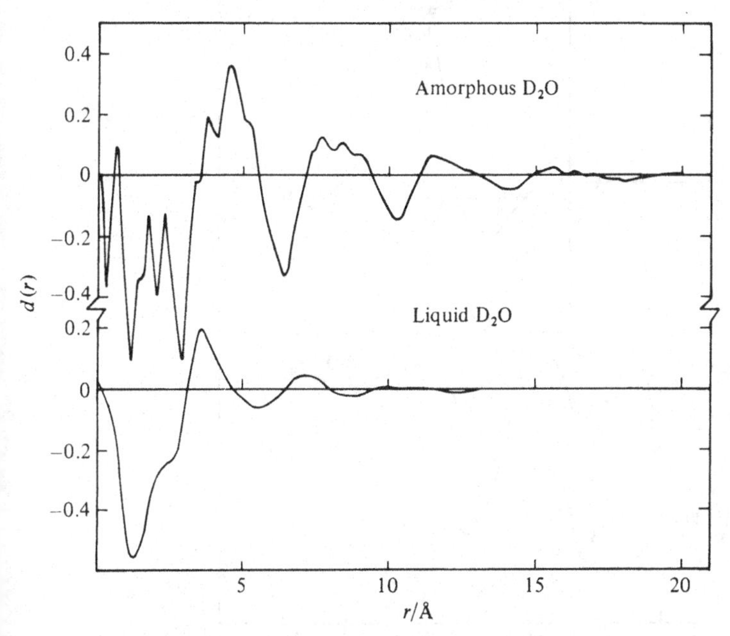

Figure 20. The $d(r)$ functions for a-D_2O and liquid D_2O obtained from figure 19.

solid and these can be directly related to a tetrahedrally coordinated cluster, as shown in the inset. The central molecule forms two acceptor and two donor bonds to the neighbouring four molecules but the two outer protons are not in any fixed positions. This basic structure corresponds to a local region of either crystalline ice Ih or Ic and is equivalent to the pentamer cluster used by Ohtomo *et al.* [**52**] shown in figure 16.

The results of figure 20 show that spatial correlations extend well beyond the range covered by the adjacent molecules. Since coordination of the oxygen atoms is approximately tetrahedral it is possible to build an extended model of the structural relationships. Fortunately the structural properties of disordered tetrahedral networks have received much attention due to the importance of amorphous semiconductors such as a-Si and a-Ge. Boutron and Alben [**79**] adapted a random network model due to Polk [**80**] in order to simulate a-H_2O by using the main sites for oxygen atoms and adding hydrogen atoms at appropriate positions on the O–O bonds. The three partial correlation functions $g_{OO}(r)$, $g_{OH}(r)$ and $g_{HH}(r)$ were then evaluated and the transformed results showed good agreement with the earlier $S_M(Q)$ data. [**77**] The more extensive data of Chowdhury *et al.* [**78**] permit a detailed comparison of the $d(r)$ functions, and these are shown below in figure 21; the

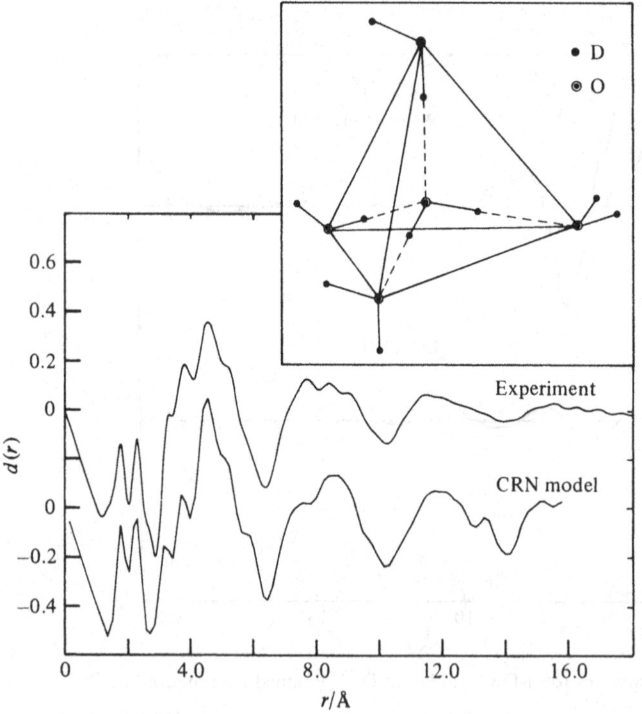

Figure 21. Comparison of the $d(r)$ function for a-D_2O with predictions from the CRN model. [**79**]; the inset shows the tetrahedral structure of hydrogen-bonded neighbours.

original $g(r)$ predictions of Boutron and Alben have been broadened by a convolution technique to give similar widths to the first two intermolecular peaks. The two curves for the experiment and the model show remarkable agreement over a range extending up to 12 Å; beyond that distance it is the accuracy of the model that breaks down, not the experimental data! The results show that the low-density form of amorphous ice may be characterized as a continuous random network based on complete connectivity of the hydrogen bonds. The topology of the network cannot be easily determined from the available diffraction data. It would be interesting to discriminate between five- and six-membered rings as discussed by Stillinger [**81**] and Chowdhury, Dore and Montague [**82**] but further experiments would be needed to isolate the partial functions in the intermediate range; this type of measurement (section 6) is feasible using existing facilities but it would be a difficult experiment to carry out. More recently, Speedy [**83**] has suggested even more complicated self-replicating structures for water on the basis of thermodynamic arguments, but there is little direct evidence for these geometries from the diffraction data.

4.5. *Other Types of Experiment*

An experiment to separate the spin-incoherent scattering from the coherent contribution has been attempted for liquid D_2O by Dore, Clarke and Wenzel [**41**] using polarization analysis techniques based on the triple-axis instrument D5 at ILL. The third scattering served as a means of analysing the spin-state of the beam scattered by the D_2O sample, and by using a spin-flipper in the incident beam it was possible to isolate the spin–flip and no-spin–flip components that would occur for the scattering of an unpolarized beam. The data were analysed in terms of eqn (3.1) and the results are shown in figure 22. It is clear from the top graph that the spin-incoherent contribution produces a smooth curve, since there can be no interference effects section 2.1) and it is the coherent scattering cross-section which resembles the structure factor for D_2O, as expected. The lower curves show a comparison of the summed components and a conventional measurement of total scattering observed in the two-axis mode using the same instrument. It is clear that the two curves diverge as the Q-value increases; this is presumably because the width of the energy resolution profile is insufficient to detect the full range of neutron energies resulting from the inelastic/quasi-elastic scattering from the water sample. Although the main objectives of the experiment have been achieved it is clear from the figures that there is a considerable spread in the points due to statistical errors. This disappointing result shows that the information obtained does not yield a more precise distribution for the incoherent scattering profile than can be obtained by calculation using an established procedure to evaluate the effects of inelasticity. Although there have been significant improvements in polarization techniques,

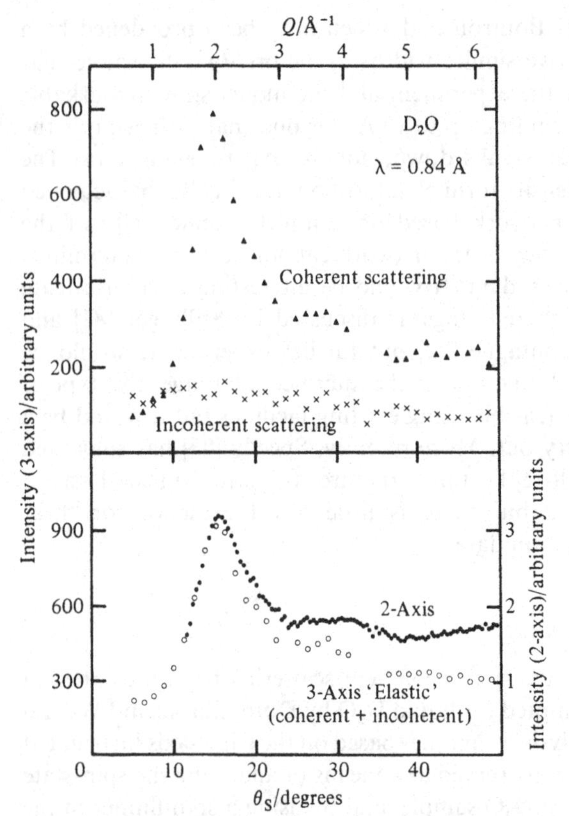

Figure 22. Separation of coherent and spin-incoherent scattering from liquid D_2O using polarization analysis techniques.

the basic limitation is due to the low count-rate resulting from the polarizing monochromator and the relative inefficiency in the use of an analyser crystal scattering for the third scattering.

As the inelasticity corrections have such a direct influence on the measured diffraction pattern, it is interesting to investigate the dynamic properties of the system by measurement of $S(Q, \omega)$. As a large Q-range is needed, the wavelength must be kept in the region of 1 Å or lower, and many conventional neutron spectrometers are not suitable because these normally use incident neutrons in the 4 Å range in order to obtain satisfactory resolution in the energy transfer $\hbar\omega$. Chowdhury, Dore and Suck [**38**] used the rotating crystal spectrometer IN4 at ILL to measure $S(Q, \omega)$ for a wide angular range with incident neutrons of 74 meV, corresponding to a wavelength of 1.05 Å. The rotating crystal spectrometer uses a time-of-flight technique to analyse the energy spectrum of the scattered neutrons. After the usual experimental corrections were applied the symmetrized dynamic structure factor $\tilde{S}(Q, \omega)$ was evaluated and is shown in figure 23; there is good agreement in the data

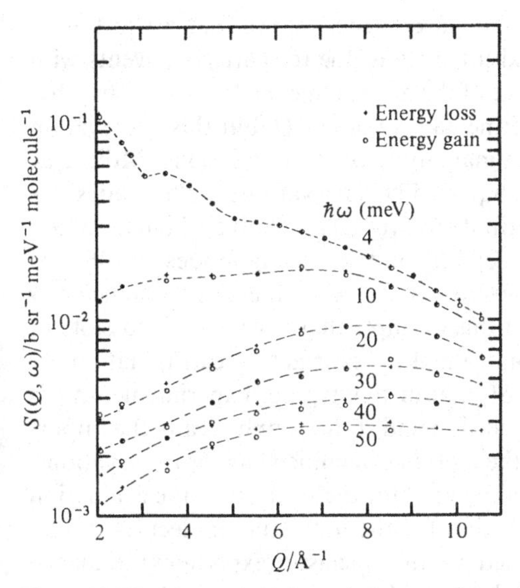

Figure 23. The symmetrized dynamic structure factor $\tilde{S}(Q,\omega)$ obtained for inelastic studies on liquid D_2O.

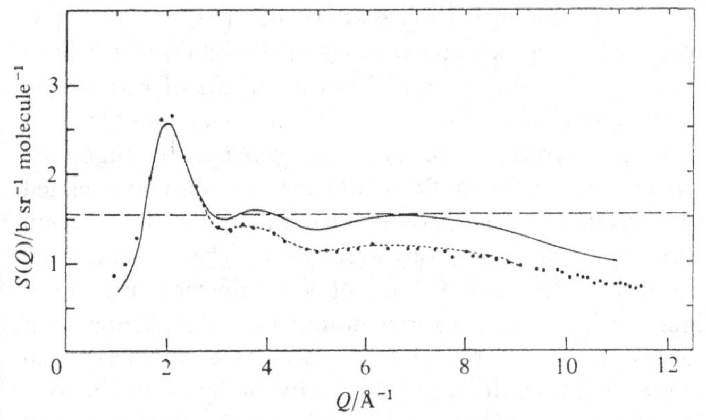

Figure 24. A comparison of the integrated intensity from the inelastic (three-axis) experiment with that from the diffraction (two-axis) experiment.

taken for energy gain and energy loss conditions, confirming the principle of detailed balance. The results extend to Q-values of 10 Å^{-1} and cover an energy transfer range of -145 to $+56$ meV. Several different procedures may be used to relate these functions to either the observed total diffraction pattern, $d\sigma/d\Omega$, or the true $S(Q)$ function obtained by direct evaluation of the zero moment, $\int S(Q,\omega)\,d\omega$. Figure 24 shows a comparison of the integrated cross-section for the energy range covered in the experiment with the total differential cross-section observed in a two-axis diffraction experiment. The overall shapes are in good agreement but there is a clear difference in the fall-off at Q-values greater than 3 $Å^{-1}$. This discrepancy shows that a substantial contribution to the total scattered intensity is due to scattering events with an energy transfer outside the range of the measurements. It is surprising that the deviations are so large at intermediate values of Q, but this observation provides a clear indication of the magnitude of the corrections. Molecular liquids, as opposed to monatomic liquids, [**39**] have various excited states due to internal motion and these contribute to $S(Q, \omega)$ at relatively high ω-values. For water there are well-defined bending and stretching modes at 146, 305 and 320 meV which will only be excited if the neutron energy is sufficiently high. It is immediately clear from this example that it will be impossible to obtain sufficient experimental data to make a satisfactory extrapolation for the determination of $S(Q)$ from $S(Q, \omega)$ in this type of experiment. This is an unfortunate but inevitable conclusion and shows emphatically the importance of the vibrational modes of the molecule in influencing the observations. It is therefore necessary to make allowance for these effects in the evaluation of the effective scattering cross-section for the individual molecules. It can be seen from the results obtained in this inelastic experiment that the calculated values are certain to be considerably more reliable than any experimentally measured quantities.

Although this review is primarily concerned with structure on the molecular scale the observation of small-angle X-ray scattering in H_2O water at low temperatures is interesting as it indicates features in the liquid which have correlations in the region of 10 Å. The SAXS measurements of Bosio *et al.* [**85**] showed a significant contribution to the scattered intensity at low Q-values (< 0.2 $Å^{-1}$) which was temperature-dependent and increased in magnitude as the sample temperature was reduced. Both H_2O and D_2O showed a related but not identical behaviour and the SAXS signal was destroyed if small quantities of ethanol (~ 0.1 mole fraction) were added. These results were interpreted in terms of the formation of transient water clusters with a local density that was different from that of the surrounding liquid. The phenomenon can be readily described in terms of percolation concepts (section 5.3) which refer to a collective correlations in the connectivity of hydrogen-bonded molecules. A further experiment [**86**] using the small-angle neutron scattering (SANS) instrument, D17, at ILL was made on D_2O water in the normal and under-cooled liquid phases but surprisingly there was no evidence of a SANS signal. This factor was attributed to the incoherent scattering contribution from the sample which makes the SANS measurements less sensitive than the SAXS studies due to the increased 'background' scattering from the bulk sample. Another possible explanation could be due to the velocity of the neutron. For X-rays the static approximation is well satisfied and the SAXS signal gives a 'snap-shot' picture of the density fluctuations, but in the case of neutrons the interaction time with the cluster is much longer and is typically 2.10^{-15} s for a diameter of 10 Å. This value is comparable with the vibrational motion of the molecules in the sample, but still seems considerably shorter

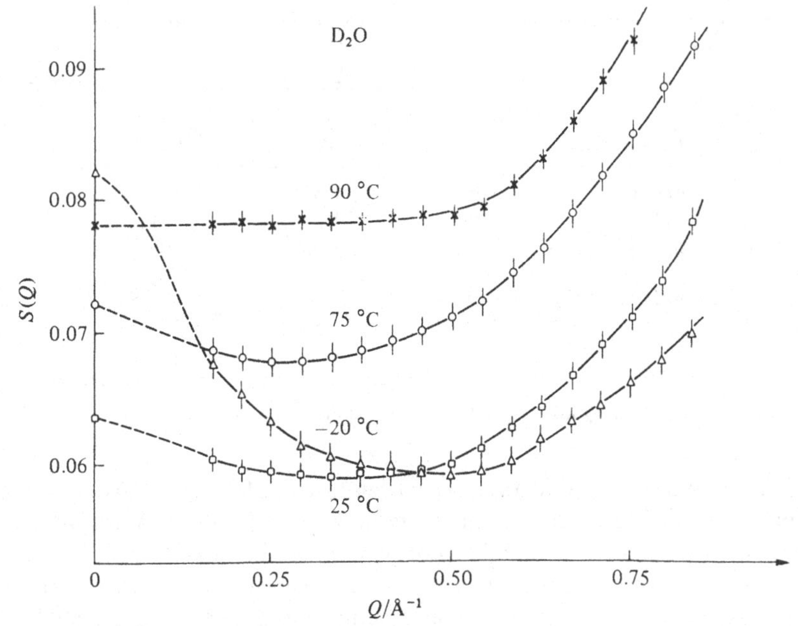

Figure 25. Small-angle X-ray scattering by liquid D_2O.

than the predicted lifetime of a hydrogen bond which is expected to fall in the range 0.1–2 ps. Further experiments using both SAXS and SANS facilities are obviously desirable.

5. Structural Information Obtained from Studies of D_2O

5.1. *Molecular Conformation in Vapour, Liquid and Solid Phases*

The geometrical conformation of a water molecule is dependent on its environment. Although it is difficult to make direct comparisons for results based on different techniques, the accepted values of the r_{OD} intramolecular bond exhibit a substantial change for different phases as shown in table 3. There is a systematic trend towards a higher value for the phases which show greater degrees of hydrogen bonding. Whalley **[87]** questioned the high values of 1.01 Å for ice Ih on the basis of an analysis of Raman spectra and other information, but further neutron measurements by Kuhs and Lehmann **[88]** confirmed this value and also provided verification of the linear geometry of the hydrogen bond. Subsequent measurements **[88]** have shown that the proton has a relatively large librational amplitude and an asymmetric distribution profile which extends along the hydrogen-bond axis towards the mid-point. The subject of the hydrogen (deuterium) position in various phases of crystalline ice remains an active topic for current research, and several

Table 3. *Values of the intramolecular distances of the water molecule for various phases*

Phase	Experiment	r_{OD} (Å)	r_{DD} (Å)	θ_{DOD} (deg)
Vapour	e/D_2O	0.96	1.515	104.5
Liquid	n/D_2O	0.98	1.55	104
Amorphous solid	n/D_2O	1.00	1.58§	109.5§
Ice Ih	n/H_2O, D_2O	1.01	—	109.5
Ice II	x/H_2O	0.983	1.543	105.4
Ice IX	x/H_2O	0.985	1.567	105.3

§ Indicates a fixed value not treated as a variable parameter.

interesting and unexpected features have been revealed [**89**] from the detailed treatment of high-precision data taken on modern instruments. Amorphous D_2O ice appears to have a slightly lower r_{OD}-value than ice-Ih, which is in keeping with the lower temperatures and bond-angle variation caused by the disorder of the network structure. The experimental data are unfortunately unable to define the DD distance sufficiently precisely to determine the mean DOD angle, which is probably close to the tetrahedral value of 109.5°. Electron diffraction studies on the isolated molecule give a value of 0.966 ± 0.003 Å for r_g, and the DOD angle is reduced to 104°. The situation in liquid water appears to be an intermediate one, in which the bond-length is larger than that of the vapour but retains a DOD angle that is less than the tetrahedral value. Since the adjacent molecules are in motion and there is no regular interlinked structure in the liquid phase it is easy to see why the angle remains similar to that of the vapour. As a consequence, the molecular conformation could change with variation in the external conditions such as temperature and pressure. However, the effects are not likely to be large, and Whalley [**87**] has predicted a rate of 10^{-4} Å kbar^{-1} from pressure effects on Raman spectra; this small variation is well beyond the accuracy of present-day techniques. It therefore seems that the conformation changes arise primarily from the geometrical factors affecting the hydrogen-bond angles. Existing neutron diffraction measurements on D_2O liquid over a temperature range of −14 to 80 °C [**46–53**] and a pressure range of 1 bar −15 kbar [**54**, **57**] have shown no evidence for changes in the overall molecular conformation, but the neutron results of Neilson [**57**] suggest a possible change in the vibrational amplitude.

5.2. *Comparison with Computer Simulation Results*

A considerable literature is devoted to the computer simulation of the properties of liquid water with molecular dynamics or Monte Carlo programs based on specified interaction potentials. The predictions from the different model descriptions can be compared with the available structural information to give some idea of whether the chosen form of the interaction is able to give a satisfactory representation of the real liquid. It is not the purpose of this review to consider the different methods in which the parameters defining the various models have been chosen as this is covered in a separate article in this volume. [**89**] Nevertheless, some comment needs to be made on the relation between the experimental data and the simulation results for D_2O; a further discussion of information on the separated partial $g(r)$ functions is deferred until section 6.

The most successful of the early formulations is the ST2 potential of Rahmann and Stillinger [**81**], which is still used extensively. Other workers prefer to use the MCY potential [**67**], which is based on *ab initio* calculations and seems to give better agreement with the structural data. Another representation with a relatively simple interaction is the TIPS potential of Jorgensen. [**91**] These and other forms of the potential are based on a classical treatment of discrete molecules using a site–site interaction with distributed charges to simulate the hydrogen bond. Other developments which introduce effective three-body interactions through polarization effects have been considered by Barnes *et al.* [**92**] In the case of the X-ray diffraction results, all predictions give a larger height to the first peak in $g_{OO}(r)$ than is obtained from the X-ray data. This means that most models create too much ordering into a local tetrahedral configuration. A similar effect is apparent with the neutron data, as shown in figure 26, for a prediction with the ST2 potential. The figure shows how the individual partial components add to give the composite function observed in neutron diffraction by D_2O. Although resolution effects (Appendix 3) can broaden the neutron $g(r)$ distribution it

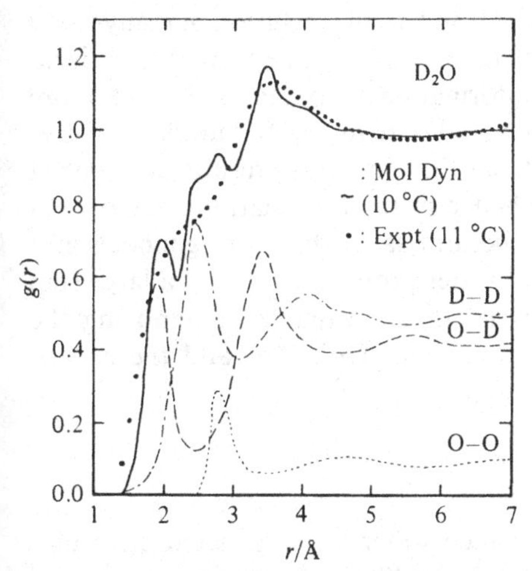

Figure 26. Combination of partial $g_{\alpha\beta}(r)$ functions from ST2 predictions with the composite $g^N(r)$ function given by neutron scattering from liquid D_2O.

is clear that the fine structure in the region of 2–3.5 Å is not seen in the experimental results. However, the general features are in good agreement and a relatively small increase in the widths of the separate peaks could lead to a much smoother composite function. In recent years it has been suggested that quantum effects due to zero-point motion would be sufficient to give the required spread, and this is a topic of much current interest. [**93, 94**] The curves show that there is substantial compensation in the composite $g(r)$ pattern for D_2O due to the fact that the oscillatory forms of the OD and DD components are of similar magnitude and out of phase. It is therefore apparent that this single neutron measurement on D_2O is not a particularly sensitive probe of the interaction potential once it is realized that the tetrahedral coordination plays a major role in the structural properties and that this is already incorporated into most successful models of the interaction potential.

The sensitivity of the structure to variations in temperature and pressure provides a much more stringent test of the effectiveness of the formulation. In this respect the work by Egelstaff, Chen and collaborators [**10**] is of relevance, since the data are compared directly with LCY predictions [**66**] using a modified form of the MCY potential. The results show the correct qualitative behaviour, but underestimate the magnitude by a factor of two as shown in figure 8. A corresponding analysis of the temperature variation for X-ray diffraction data by Egelstaff and Root [**95**] shows a similar behaviour, and they suggest that the discrepancies between theory and experiment in both pressure and temperature derivatives are closely related and also indicate the non-additivity of the interaction forces. A method of introducing a pseudo-potential to correct for an apparent temperature shift between 'real' water and 'computer' water is proposed. One factor to emerge clearly from the wide range of experimental information presented in the previous sections is that collective effects are expected to play a significant role in the description of the real liquid. However, the inclusion of many-body terms creates serious difficulties for the theoretical treatment since there is no reliable way of obtaining accurate information to parameterize the more complicated interaction potential that would be required. It is not practicable, at present, to evaluate these effects from the diffraction results although it is an obvious prerequisite that any change in the formulation should yield better agreement with the available structural data. The existing models give a satisfactory first-order description of water properties but it is a large step to introduce a more 'realistic' potential that is capable of removing the observed discrepancies between the computer predictions and the experimental results.

5.3. *Percolation Properties*

The anomalous properties of under-cooled water have attracted particular attention in recent years. Stanley and Teixeira [**96**] have introduced a model in which the hydrogen-bond connectivity plays a major role in the description of the system. It was shown that the anomalous behaviour of the isothermal compressibility K_T, and the specific heat C_P, could be represented in terms of a single parameter p_B, which governs the probability of a hydrogen bond between adjacent molecules and increases as the temperature decreases. The correlated site percolation model shows that clusters of hydrogen-bonded molecules are formed and that the sizes are dependent on p_B and hence, T. Further studies and comparison with predictions for ST2 water [**97**] showed that the statistics of the hydrogen bonding was not dependent on the energy or geometric criteria chosen to define whether a hydrogen bond had been formed but was an inherent property of the network structure. If p_B becomes large there are significant numbers of four-bonded molecules in close proximity, so that a hydrogen-bonded cluster of finite size is produced. Since this is presumed to have the open tetrahedral structure of crystalline or amorphous ice, it creates a region of low density relative to that of the rest of the liquid and is therefore responsible for enhanced density fluctuations in the condensed phase. The observation of a temperature-dependent small-angle X-ray scattering (SAXS) signal in under-cooled water [**84**] provides direct support for this concept of transient cluster formation (section 5.2).

The percolation or gelation model is formally established as a topological scheme which describes the statistical nature of the bonds rather than the sites. It cannot therefore be converted to an equivalent spatial representation in real space. The fact that the computer simulations by Geiger and collaborators [**100**] support this treatment indicates that it is a satisfactory alternative way of describing the statistical nature of the hydrogen bonding and that p_B is essential in a formal description of the fundamental properties.

5.4. *Spatial Correlations from Neutron Studies of D_2O*

Although only limited information can be deduced from neutron studies of D_2O alone, a number of important conclusions can be established. A comparison of the basic $d_L(r)$ curves for the liquid with those of the amorphous solid given in figure 2 emphasizes that the spatial correlations extend over a wider range of r-values. It is not possible to discriminate between a static disorder due to the local hydrogen-bonded configurations or a dynamic disorder due to the constant relative motion of the molecules but it is clear that the discrepancy with the computer predictions does not arise from instrument resolution effects. The trends seen in the temperature variation studies do appear to support Rice's conjecture that the amorphous ice network provides a first-order conceptual framework for regarding the spatial distribution. In this sense, low-temperature water can be regarded as either a network structure with defects or a continuous structure with a large variation in the hydrogen-bond bend angles. The temperature variation studies confirm this basic concept, but as shown in figure 27 suggest that under-cooled water at 14 °C is still very far from a structure like amorphous

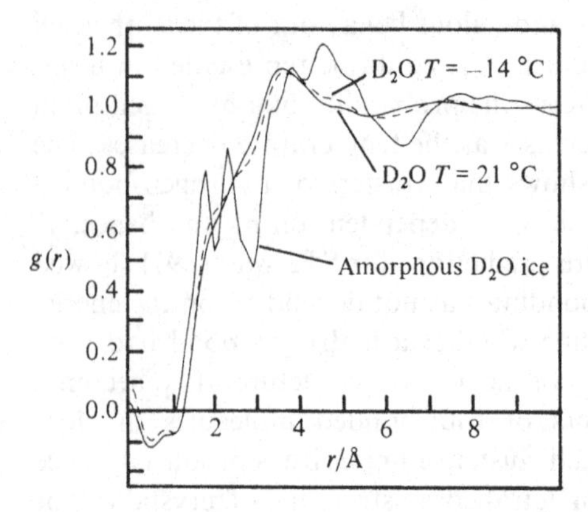

Figure 27. Comparison of the $g(r)$ function for neutron scattering by D_2O in the normal liquid, under-cooled liquid and amorphous solid phases.

ice. It seems that this picture is also a useful aid to the discussion of characteristics observed in spectroscopic investigations. [**101**] The increase in slope $(\partial Q_0/\partial T)$ corresponding to the change in peak position with temperature shown in figure 10 implies a more rapid change of structure as the temperature is reduced, but this is partly due to the compensating reduction in density; this feature is discussed more fully in section 6.6. Another possibility is to consider alternative network structures, and Hajdu [**101**] has suggested a form based on a tetragonal lattice structure which gives good agreement with X-ray diffraction results for the liquid. It would be interesting to see if this modified local structure – which is effectively an asymmetric distortion from tetrahedral symmetry – had any advantage in describing the neutron data, but this computation has not yet been made.

The structural properties of amorphous ice itself may prove to be rather more variable than at first thought although it now seems that the low-density form is the one most commonly formed by vapour deposition. The photographic X-ray patterns obtained by Mishima *et al.* [**74**] for amorphous ice prepared by application of pressure to crystalline ice below the glass transition temperature showed substantial changes in the shape and position of the main diffraction peak according to the methods used. The recent electron diffraction studies of water clusters consisting of 20–200 molecules by Torchet *et al.* [**102**] are also of interest in this context since the Boutron and Alben model [**79**] did not fit the data and an alternative structure based on distorted rings of three to six molecules was proposed. However, it may be inappropriate to cite these experiments as evidence against the continuum-random network description since the materials formed are probably unable to adopt a minimum-energy configuration. In the case of clusters it is also possible that surface effects play a significant role in the structural relationships.

The nature of the hydrogen bonding in the normal liquid phase is more difficult to characterize without further information. The pentamer/monomer mixture model of Ohtomo *et al.* [**52**] is instructive in providing a convenient parameterization of the main trends in the diffraction pattern. However, the x_5 values are generally much higher than those given by spectroscopic techniques, and since x_5 goes to unity at a temperature of 15 °C this gives little latitude for the description of the substantial changes which occur at lower temperatures. Since the model assumes no orientational correlation between pentamer units this reinforces the view expressed previously that there are significant changes in correlation across second-neighbours. These limitations are recognized in a later paper [**103**] which considers the partial structure factors. It would presumably be possible to introduce a formalism in which a temperature-dependent local correlation could be introduced on a hierarchical basis of larger polymeric units, but it is doubtful that this would lead to a more effective description than the network system previously discussed.

5.5. *The Relationship of Water to Other Hydrogen-Bonded Liquids*

The problems addressed in the previous sections could equally well have been considered in relation to other hydrogen-bonded systems. Although extensive neutron studies of hydrogen bonding in crystalline materials have been reported there is little corresponding information on the liquid phase even for relatively simple molecules. The currently available neutron data are restricted to ammonia, hydrogen halides and short-chain alcohols. It is not appropriate to give a detailed appraisal of these results in the current review, but a general comparison reveals a number of interesting facts. Neutron studies of liquid ND_3 and CD_3OD have been reported over a range of temperatures, and the vapour-deposited solid phase has also been investigated. The main diffraction peak, Q_0, is found to vary in a systematic way as shown in figure 28, which also gives the corresponding variation of densities. If all distances can be scaled according to the density, then the product $Q_0\rho^{\frac{1}{3}}$ should remain constant. This relation is surprisingly well obeyed by the ND_3 and CD_3OD results and the curves also appear to extrapolate smoothly to the values for the amorphous phase at a temperature known to be close to the glass transition temperature, T_g. The behaviour of water seems to be quite different due to the anomalous density variation in the under-cooled region. However, it does contain a well-defined systematic trend which possibly extrapolates to the Q_0-value of 1.7 Å^{-1} for a-D_2O in the region of −40 °C, which is similar to the Angell/Speedy temperature, T_A, that characterizes the divergence observed in various physical properties. [**104, 105**] These features imply that under-cooled liquid water may be evolving towards the complete network structure of hydrogen bonds exhibited by amorphous ice as the temperature is reduced into the more deeply under-cooled region. This viewpoint is entirely in accord with the principles of percolation presented

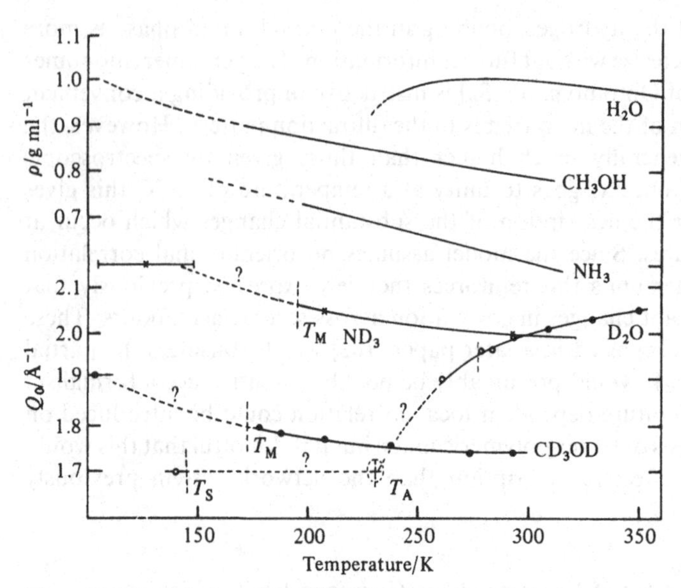

Figure 28. The variation of the diffraction peak, Q_0 and the density ρ for various molecules in the liquid and amorphous phase.

in the previous section, but further experiments [**106**] will be required to confirm that this behaviour persists over the extrapolated temperature range. The production of the low-density form of vitreous ice directly from the liquid [**72, 73**] is also a strong indication that this is probably the natural form of disordered ice produced when homogenous nucleation is inhibited. However, the earlier controversy [**107**] over the continuity of state which was based on thermodynamic arguments is still not fully resolved.

The deuterium (hydrogen) halides form an important class of hydrogen-bonded liquids that deserve a full-scale review to themselves. The neutron diffraction studies give $d(r)$ functions that exhibit well-defined hydrogen-bond peaks, although there is some evidence of asymmetric distortion of the intramolecular bond. [**108, 109**] The most interesting results have been obtained for DF [**109**] and a single HF/DF mixture [**110**], where there is strong evidence for the formation of hexamer rings. As these molecules act as a single donor/single acceptor species, the hydrogen-bonding conformation can only produce chains or rings. Hydrogen fluoride is one of the strongest hydrogen-bonding molecules known, and the data suggest that there could be strong collective effects influencing the ring structure. Under these conditions the identity of the individual molecules is lost in the collective resonance behaviour around the ring. It is clear that much more work on the precise determination of spatial correlations for intra- and intermolecular hydrogen-bond distances is required. The tetrahedral nature of the double donor/double acceptor interaction for the water molecule seems to be an important criterion because it permits a space-filling network to form rather than a ring or chain. The ammonia molecule has a triple donor/triple acceptor interaction and the crystalline form exhibits six bent hydrogen bonds for each molecule. The geometrical constraints therefore make it more difficult to form a satisfactory space-filling network of linear hydrogen bonds and a compromise in the orientational ordering required for production of a stable configuration. No structural prediction seems to have been made for disordered networks with sixfold coordination.

6. Investigation of Partial Pair Correlation Functions

Although X-ray studies of H_2O and D_2O coupled with neutron studies of D_2O provide important information, the two composite pair-correlation functions derived from these two data sets do not permit a separation into the three partial pair correlation functions required for a more complete description of the liquid structure. In order to achieve this separation, a third independent diffraction measurement is needed. The first attempt to determine the three partial functions was made by Palinkas, Kalman and Lengyel **[111]** using the results of electron diffraction **[112]** to provide a third measurement. Since that time several groups have used hydrogen/deuterium isotope substitution with neutron diffraction to obtain the same information. Unfortunately there are substantial differences in the conclusions reached by these groups and a clear picture has not yet emerged. The various experimental measurements on the H_2O/D_2O mixtures are given in table 5 in section 6.2. The following sections provide a survey of the available data and include a critical assessment of the procedures used in the collection and analysis of the experimental measurements.

6.1. *Partial g(r) Data of Palinkas* et al.

The use of electron diffraction for the structural investigation of the condensed phase is more familiar in the context of LEED studies of surfaces by reflection geometry or defect studies in solids by transmission geometry. The high scattering power of the samples means that only small thicknesses of material can be placed in the main beam if a satisfactory diffraction pattern is to be observed. This situation provides adverse experimental conditions for the study of liquids. However, Kalman, Palinkas and Kovacs **[112]** have overcome these experimental difficulties by making electron diffraction measurements on thin films of water formed inside the evacuated chamber of the diffractometer. The films are typically 1000 Å thick and last for only a short time, but the scattering is sufficiently intense to record the diffraction pattern on a photographic plate within a few seconds. The processed plate therefore can be used to give the required intensity information, which is analysed using standard electron diffraction methods. The formalism for defining the scattering profile contains terms from both nuclear and electron contributions and therefore the form-factors are more complicated than for either X-rays or neutrons. The measurements were conducted on both H_2O and D_2O liquids at a temperature of 5 °C, but details of the analysis were presented only for D_2O, since the data could be combined with corresponding information from X-ray and neutron studies. The intramolecular r_{OD} value was found to be 0.96 Å, which is comparable with that in the gas phase and a little shorter than the value given by recent neutron experiments on the liquid (table 3). The hydrogen-bond distance of 1.95 Å seems large, but the broad peak also contains an unresolved component from the intramolecular DD contribution; the various peak positions suggested that the hydrogen-bond was non-linear. In a second paper by Palinkas, Kalman and Kovacs [**111**] the electron diffraction results are combined with available X-ray data [**18**] at 5 °C and neutron data [**45**] at 25 °C, in order to separate the three $g_{OO}(r)$, $g_{OD}(r)$ and $g_{DD}(r)$ functions. This is an involved process due to the different 'atomic' and 'nuclear' form-factors which apply to each of the experimental data sets, and the various assumptions must be made in the computation of the combinations; these details are given in the paper.

The results obtained by Palinkas *et al.* [**111**] for the partial functions are shown in figure 29 as a composite plot (eqn (2.12)) with the contributions from each term weighted according to the factors for neutron scattering by D_2O; the experimental results for neutron studies of D_2O by Gibson [**49, 50**] are also superimposed and can be compared with the corresponding values of Narten [**45**] used in the original analysis. The results at low r-values show that there is considerable structure due to the first OD and DD peaks, which adds to give a smooth curve for the composite neutron/D_2O results. This

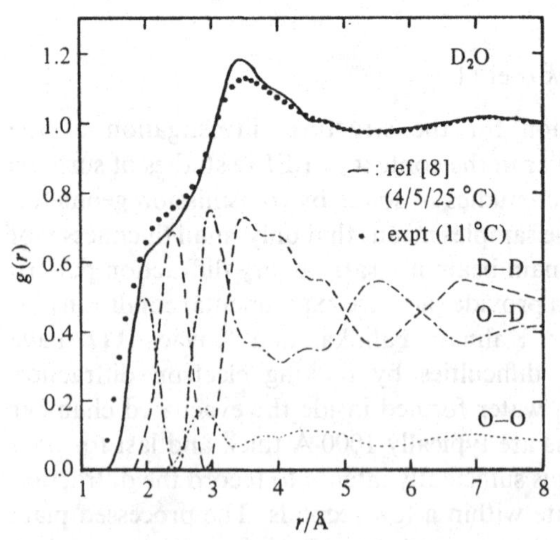

Figure 29. Partial pair correlation functions for liquid D_2O compared with computer simulation results.

behaviour is similar to that found in the computer simulations with the ST2 potential (figure 26), except that the measured peaks are broader and the sharp features seen in the computer predictions for the composite curve are no longer revealed. The broadening due to instrumental resolution functions (section 3) is insufficient to account for the discrepancies. These results confirm the basic tetrahedral nature of the hydrogen bonding and emphasize that the apparently featureless curve obtained from the neutron scattering data for D_2O gives a somewhat misleading impression about the local correlations, which are much more structured than is initially apparent.

At larger r-values (> 5 Å) there is a similar behaviour in the experimental results where the $g_{OD}(r)$ and $g_{DD}(r)$ have a similar oscillatory structure that is out of phase. In this region the computer predictions show little evidence of an oscillatory behaviour. Since the separate neutron and X-ray data show only small oscillations, it is clear that this phenomenon can only originate from the solution of the three simultaneous equations if the electron diffraction results show significant structure in this region. This possibility raises the question whether the inverted behaviour of the two functions in an artefact arising from systematic errors in the treatment of the three data sets. [**113**] Since three radiation techniques with different self-scattering form-factors are used there would seem to be scope for discrepancies of this type to enter the treatment of the data. This factor, the apparent discrepancy with computer predictions, and the fact that the data sets were not taken at the same temperature has caused many people to disregard the conclusions presented by this analysis, partly due to a mistrust in the accuracy of the electron diffraction results. The possibilities of systematic errors will be discussed in a later section.

6.2. *Partial g(r) Data from Reactor Neutrons*

There is clearly a disadvantage in combining diffraction data taken on different instruments in order to separate the three partial functions, since systematic uncertainties and instrumental differences such as resolution and sample geometry can contribute to errors in the final results. The use of isotopic substitution in neutron diffraction experiments is therefore very attractive in this case, particularly as the D and H isotopes have such widely differing coherent scattering lengths of $+6.67$ fm and -3.74 fm, respectively (table 1). The large incoherent scattering of hydrogen remains the main obstacle to be overcome if satisfactory results are to be obtained.

The first attempt to exploit the variable b-values of deuterium and hydrogen was made by Powles, Dore and Page. [**114**] Neutron scattering was measured for a thin (0.2 mm thick) planar sample of H_2O and a special mixture of H_2O/D_2O. If hydrogen and deuterium are assumed to be equivalent then the effective b-value for the mixtures may be written as

$$\langle b_{HD} \rangle = \alpha_D b_D + (1-\alpha_D) b_H, \tag{6.1}$$

where α_D is the fractional concentration of deuterium atoms in the mixture. Since b_H and b_D are of opposite sign, it is possible to choose a value of α_D which gives an effective zero value for $\langle b_{HD} \rangle$. Under these circumstances there is complete cancellation of the coherent scattering from the combined assembly of hydrogen and deuterium atoms and a corresponding increase in the isotopic incoherent scattering. These conditions are created when $\alpha_D = 0.359$, and the interference effects in the cross-section for this H_2O/D_2O mixture will be solely due to the distribution of the oxygen atoms and should therefore resemble the X-ray diffraction pattern. The results of the measurement are shown in figure 30 and confirm the basic principles, but also indicate the very large contribution which arises from both spin and isotopic incoherence. As a result of the extreme inelasticity effects the cross-section falls off rapidly with scattering angle. The structural information is to be obtained from the small oscillatory pattern superimposed on the overall curve, and therefore requires both a high statistical accuracy and an accurate evaluation of the fall-off. In this case the known structure factor from the X-ray studies was used to determine the shape of the incoherent scattering profile in order to analyse the data take on pure H_2O during the same experimental run. These data are shown in figure 31 with the incoherent contribution evaluated from various assumptions. It is clear that the features of the interference terms are very different from those exhibited by D_2O.

The principles involved in the H/D substitution method mean that the effective $\langle b_{HD} \rangle$ value can be varied continuously between the values b_H and

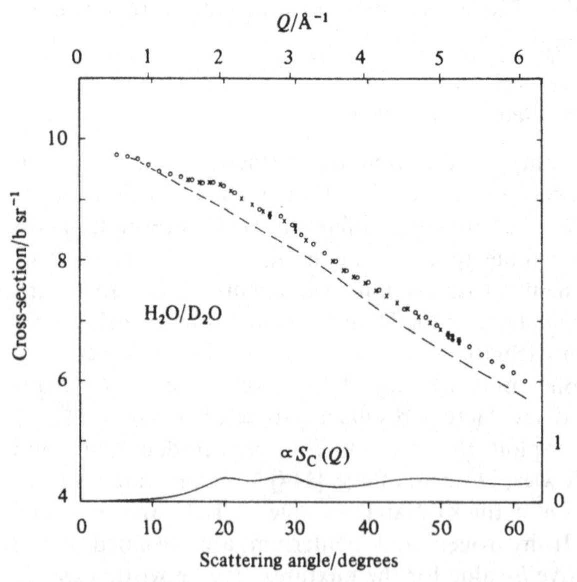

Figure 30. The differential cross-section for neutron scattering by an H_2O/D_2O mixture ($\alpha_D = 0.392$).

Table 4. *Experimental studies of H_2O/D_2O mixtures by neutron diffraction*

	Ref.	λ (Å)	α_D values	Separation $g_{OH}(r)$	Separation $g_{HH}(r)$
Narten & Thiessen	116	0.9	0.98, 0.68, 0.36, 0.01	✓	✓
Reed & Dore	115	0.70	0.98, 0.72, 0.36, 0.01	✓	✓
Soper & Silver	63	(Var.) $\theta = 40.$	0.98, 0.50, 0.01	—	✓

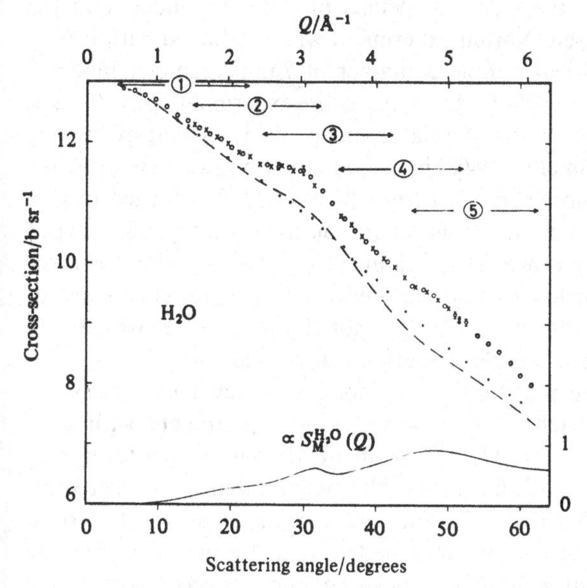

Figure 31. The differential cross-section for neutron scattering by H_2O; see text for details.

b_D. Several groups (table 4) have used this method in recent high-precision measurements. Since b_D is approximately twice the magnitude of b_H it is convenient to make a total of four measurements with approximately equal incremental changes as given in table 5. This scheme has been adopted in experiments by two groups using high-flux reactor facilities to obtain an improved statistical accuracy. Reed and Dore [**115**] have used the D4 diffractometer at ILL and Thiessen and Narten [**116**] have used the liquids diffractometer at Oak Ridge. Although similar intentions motivated the work, the experimental procedure and data analysis in these two studies show some important variations which are probably responsible for the different final conclusions. Because of the crucial nature of these difficult measurements it

Table 5. *Four 'ideal' H/D combinations for isotopic substitution experiments on water*

Sample	α_D	$\langle b_{HD} \rangle$	Weighting factors (relative)		
			OO	OD	DD
D_2O	1	6.67	0.092	0.422	0.486
D_2O/H_2O	0.718	+3.74	0.191	0.492	0.317
D_2O/H_2O	0.359	0	1	0	0
H_2O	0	−3.74	0.191	−0.492	0.317

is important to examine the technical details in order to understand the discrepancies. The Thiessen/Narten experiment was conducted with 0.89 Å neutrons, using a cylindrical sample container of 7 mm diameter made of vanadium and maintained at 25 °C. The Reed/Dore experiment at 20 °C used 0.7 Å neutrons with a planar sample based on the development of the one used in the earlier experiment. [**104**] The vanadium windows were attached to two frames with a spacer between them which could be adjusted to give variable thickness liquid samples. This is particularly advantageous in cases where the total scattering power of the sample can vary, because it enables the thickness to be adjusted so that the multiple-scattering effects are of similar magnitude for each sample and are not too large. In most neutron diffraction experiments the sample scattering is arranged to be 0.10–0.15 of the incident beam, but the need for high statistical accuracy means that some compromise must be reached (Appendix 3) for experiments with high hydrogen content. In the Reed/Dore experiments the sample scattering was approximately constant at 0.2, but in the Thiessen/Narten experiments this factor varied between 0.2 and 0.7. The detector efficiency must also be maintained constant to an accuracy of better than 0.1 per cent over the duration of the experiment, and various consistency checks need to be applied to the raw data in order to verify that this condition is satisfied otherwise the results are of no value. In the ILL experiments it required three separate attempts before results were obtained which met these stringent conditions. The corrections for container scattering are relatively unimportant in these experiments, and the accuracy of the final cross-section data is primarily dependent on the statistics of the total count and systematic uncertainties resulting from instrumental instabilities (mainly variable detector efficiency) and experimental corrections (mainly self-attenuation and multiple scattering). The final cross-section data given by the two groups are shown in figures 32 and 33. The overall features are similar and emphasize the extreme nature of the inelasticity effects on the self-scattering terms for hydrogen by the very large fall-off in the intensity with Q-value (and scattering angle). The change in shape of the superimposed oscillatory behaviour with variation of the H/D

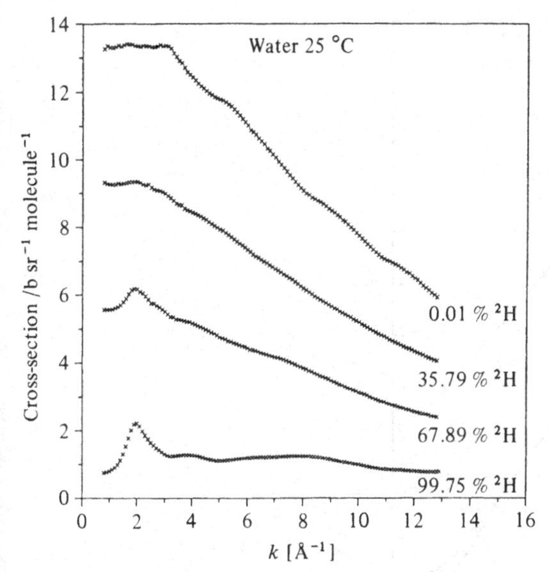

Figure 32. The differential cross-section for neutron scattering by various H_2O/D_2O mixtures obtained by the OR group. **[116]**

composition is also clearly indicated. Even at this stage in the data treatment, there appear to be some significant differences in the magnitude of the oscillations displayed by the various samples for each group. The most obvious variation is in the measurements for H_2O, where the Oak Ridge data have a well-defined oscillatory behaviour over the whole Q-range but the UKC data are much smoother, particularly in the higher Q-value region. In the lower Q-value range this trend seems to be reversed for the two water mixtures, since the oscillatory amplitudes are larger in the UKC data than in the Oak Ridge data. A quantitative comparison of the data sets and a consideration of systematic errors in the application of the associated corrections are obviously desirable but have not yet been carried out. **[117]** It seems likely that the source of the discrepancies and the variation in interpretation which follows later in the analysis has its source in the experimental criteria rather than the treatment of the analytic corrections.

The procedure required to extract the intermolecular function $D_M(Q)$ for the mixtures from the cross-section data is identical, in principle, to that for the analysis of the D_2O data. The first step is to characterize the shape of the self-scattering distribution. The introduction of hydrogen into the sample increases the inelasticity effects and gives a more rapid fall-off in the cross-section with increasing Q-value. In the analysis of the Oak Ridge data the correction terms were calculated for an asymmetric top rotator (Appendix 2) and incorporated into calculations of the cross-section for self and intramolecular contributions to the cross-section. The detailed fit to the

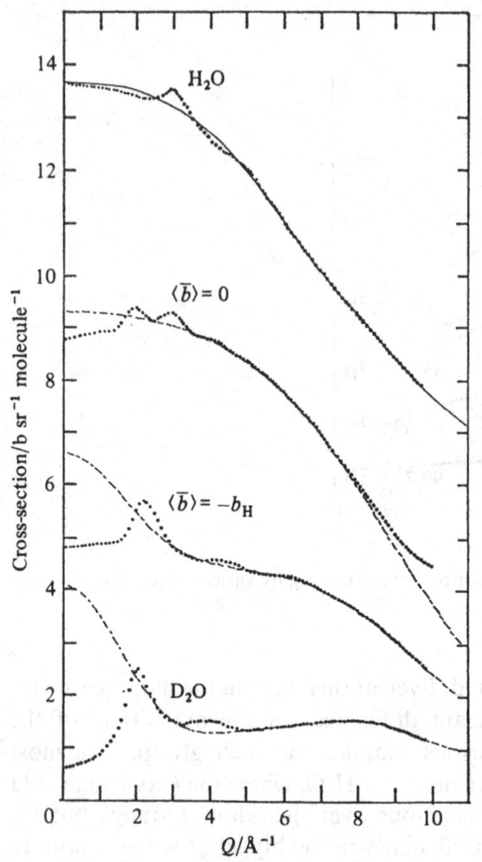

Figure 33. The differential cross-section for neutron scattering by various H_2O/D_2O mixtures obtained by the UKC group. [**115**]

overall fall-off was refined by introducing a parabolic correction term and an iteration procedure to optimize the parameters. In the analysis of the UKC data, the Powless formalism (Appendix 2) was used for the inelasticity corrections, incorporating a detailed treatment of the vibrational terms for the separate H_2O, D_2O and HDO species in the mixtures. The full expression for molecular recoil corrections was included in the treatment of the intramolecular terms and a small variation in the effective mass of the hydrogen atom was made to give agreement in the high Q-value region; the broken lines in figure 33 represent the calculated intramolecular scattering distribution for these results. In both cases the subtraction of the molecular scattering is satisfactorily carried out and, although different methods have been adopted, it is unlikely that this process will lead to significant errors (Appendix 3). The procedure is best compared with the general treatment of electron diffraction data [**118**], where it is impossible to evaluate the self-scattering component with adequate precision and the interference contribution is simply obtained by subtracting a smoothly varying function from the overall intensity pattern. In the neutron case the shape of the distribution is accessible to calculation if various assumptions are made and is effectively governed by a defined formula which can be adjusted by variation of several parameter values.

In the evaluation of the partial functions the two groups again diverge in their approach. Narten and Thiessen adopt the conventional method of solving the simultaneous equations directly to give the three partial structure factors $S_{OO}(Q)$, $S_{OH}(Q)$ and $S_{HH}(Q)$, which are then transformed over the full Q-range (< 10 Å^{-1}) to give the corresponding spatial functions $g_{OO}(r)$, $g_{OH}(r)$ and $g_{HH}(r)$. The four data sets were found to be in satisfactory agreement and produced the $g(r)$ curves **[19]** shown in figure 34. Reed and Dore adopted the unusual procedure of truncating the data at 6 Å^{-1}, transforming the composite $D_M(Q)$ functions and solving the simultaneous equations in r-space. These two methods are, in principle, identical but they involve different kinds of check mechanisms. It is clear that minor variations in the absolute scale of each intermolecular distribution function will lead to errors in the separation of the partial functions. The resulting data are therefore very sensitive to the relative scales of each interference contribution and this is directly linked to the treatment of the multiple scattering

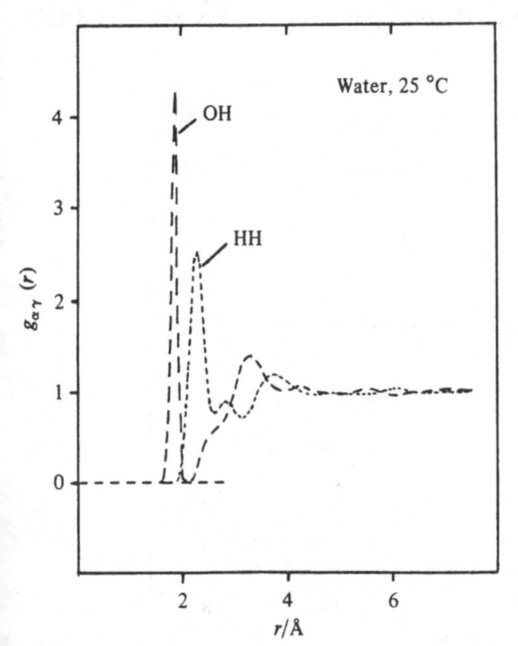

Figure 34. The partial pair correlation functions $g(r)$ obtained from figure 27 by the OR group.

(section 3). The UKC analysis therefore permits an important check on the relative scales through the behaviour of the $d(r)$ composite functions at low r-values which should follow the appropriate density line. It can be seen from figure 35 that this agreement is well established for D_2O and the two H_2O/D_2O mixtures; the negative $\langle b_{HD} \rangle$ value is responsible for the different initial slope of the H_2O data which is not susceptible to the same checks. The penalty for this procedure is a loss of resolution in the $g(r)$ functions due to the reduction of Q_m.

The final results for two of the partial functions from the Oak Ridge data [**12**] are given in figure 34. They show large variations from the partial $g(r)$ functions of Palinkas *et al.* [**111**] given in figure 29. The most obvious feature is the large initial peak in $g_{OH}(r)$, indicating a very sharp hydrogen-bond distance. Both data sets also indicate that this peak is separated from the rest of the distribution since the following dip gives $g(r)$ values that are almost zero; in this sense, both data sets disagree with the computer simulations. The initial peak in $g_{DD}(r)$ is also sharper in the Oak Ridge results although the position at 2.3 Å is in good agreement with that of Palinkas *et al.* and considerably smaller than that of the computer predictions. The experimental data for the short distances therefore exhibit similar features but the Oak Ridge results have much sharper peaks. Beyond 2.5 Å the curves have substantially different shapes. The deep minimum in $g_{DD}(r)$ obtained by Palinkas *et al.* is very much reduced in the Oak Ridge data, and the large 'out-of-phase' oscillations in $g_{OD}(r)$ and $g_{DD}(r)$ for $r > 5$ Å are completely missing.

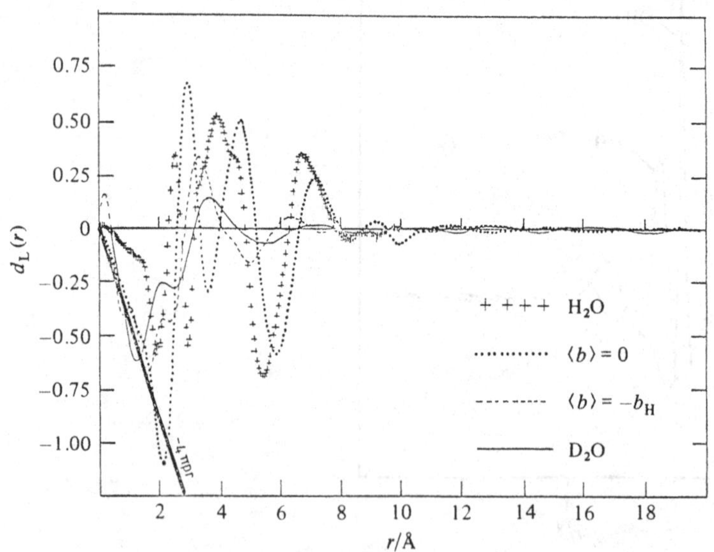

Figure 35. Composite $d(r)$ curves for H_2O/D_2O mixtures obtained from figure 33 by the UKC group.

In the analysis of the four composite data sets from the UKC experiments it was found that an unambiguous reduction to the three partial $g(r)$ functions could not be obtained from a simple analysis. This situation can arise from two sources due to either systematic errors in the data or the non-equivalence of H and D in the isotopic substitution; this latter point is discussed more fully in section 7. There are four possible ways of combining the data sets, and these were used to check the variations. The results for $g_{OH}(r)$ were found to be in excellent agreement (figure 36) if the curve for one of the ill-conditioned combinations [**115**] was deleted. The curve shows surprisingly good agreement with that of Palinkas *et al.* [**111**], including a significant oscillation amplitude in the controversial 5–8 Å region. The results for $g_{HH}(r)$ given in figure 36 show much more variation for the different combinations although there is reasonable agreement for the height and position of the first peak. In order to investigate the discrepancies more fully the separate contributions from H and D components in each data set were analysed and are shown in table 6. The interesting factor to emerge from these figures is that the related H/D proportions remain almost identical for the various combinations used in the $g_{OH}(r)$ separation but differ considerably for the $g_{HH}(r)$ evaluation. A closer examination of the different curves in figures 36 and 37 suggests that there is a systematic difference in the variation of the H/D proportions, but this type of detailed scrutiny should be viewed cautiously due to the limits in the accuracy of the data. The analysis of the UKC data therefore casts doubt on the equivalence of the H and D atoms and indicates that the simple treatment based on eqns (2.11) and (6.1) is not strictly appropriate. This additional complexity cannot be easily overcome without further measure-

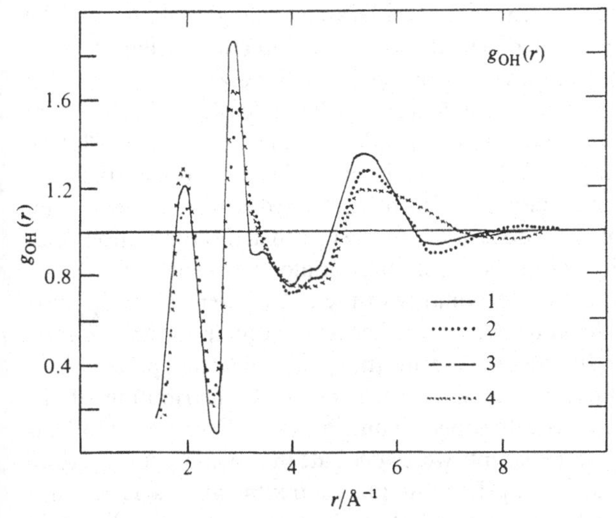

Figure 36. The $g_{OH}(r)$ functions obtained by the UKC group for different combinations of data sets; the ill-conditioned combination 3 has been deleted.

Table 6. *Separation of H and D correlations*

	Combination	Percentage contribution				
$g_{OH}(r)$	(UKC)	OD	OH	Ratio OD/OH		
	1	0.632	0.368	1.72		
	2	0.634	0.365	1.78		
	(3)	0.637	0.363	1.76		
	4	0.625	0.625	1.66		
	Combination	Percentage contribution			Ratio	
$g_{HH}(r)$	(UKC)	DD	HD	HH	DD/HH	HD/HH
	1	0.271	0.541	0.188	1.5	2.9
	2	0.413	0.453	0.134	8.1	3.4
	3	0.700	0.337	−0.037	18.3	8.8
	4	0.336	0.584	0.584	4.2	7.3
[The OR data will have similar relative weighting factors.]						
	Combination				Ratio	
$g_{HH}(r)$	(WNR/LA)	D_2O	H_2O		D_2O/H_2O	
	1§	1.904	0.460		3.57	
		DD	HD	HH	DD/HH	HD/HH
	1†	0.410	0.460	0.130	3.15	3.54

§ Quoted in reference 63.
† Re-evaluted with HD exchange terms.

ments, but the present treatment of the results appears to give support to the results of Palinkas *et al.* rather than those of Thiessen and Narten. A more detailed discussion of these features is given in section 6.4.

Another consistency check is feasible through the evaluation of $g_{OO}(r)$ from the neutron data. The special case of $\langle b_{HD} \rangle = 0$ concerns the cancellation of the coherent scattering from the H/D mixture and is therefore particularly sensitive to any differences and can also be compared with the X-ray data. Unfortunately this procedure does not yield the intended discrimination since there is some ambiguity about the most appropriate treatment of the X-ray data, which are dependent on the validity of the independent-atom approximation and can be defined in terms of either a 'molecular' or an 'atomic' distribution formalism. In order to describe the X-ray diffraction pattern fully it is necessary to evaluate the detailed electron-density distribution of the molecule and to include the hydrogen contributions. Egelstaff **[120]** has pointed out that previous treatments were adequate in providing information on the overall behaviour, but in the more precise treatment now required it will be necessary to take account of these additional factors. In these circumstances it seems that both the Oak Ridge and the UKC data are in satisfactory agreement with the X-ray results but a critical examination of the differences is obviously desirable. The main discrepancies appear to be in the relative heights of the first two peaks of $D_M(Q)$ and comparison with the X-ray diffraction pattern, but no attempt has yet been made to assess the effects in a quantitative manner. The conflicting picture emerging from these neutron experiments is discussed further in 6.4.

6.3. *Partial g(r) Data from Pulsed Neutrons*

Soper and Silver [63] have used a novel adaptation of the difference technique to isolate the $g_{HH}(r)$ partial function using pulsed neutrons from the General Purpose Diffractometer at WNR, Los Alamos. Experimental measurements were made on D_2O, H_2O and an equal mixture of H_2O/D_2O (i.e. $\alpha_D = 0.5$). The three data sets were combined directly to give the contributions for the HH interference terms. The samples were of planar form, held in an aluminium container, and the thickness was adjusted to keep the multiple scattering contribution below 25 per cent. Although data were recorded for several angles, the measurements with a 40° detector which covered a Q-range of 1.4 to 20 Å^{-1} were used in the analysis and a satisfactory difference function obtained for the 1.4 to 12 Å^{-1} range, as shown in figure 38. Due to the instrument characteristics discussed in section 4, the low-Q measurements on a pulsed source instrument have relatively low statistics and therefore the spread in the points is larger than that of the reactor studies presented in the previous section. The $g_{HH}(r)$ plot shown in figure 39 is compared with computer simulations using the ST2 and LCY potentials. The experimental

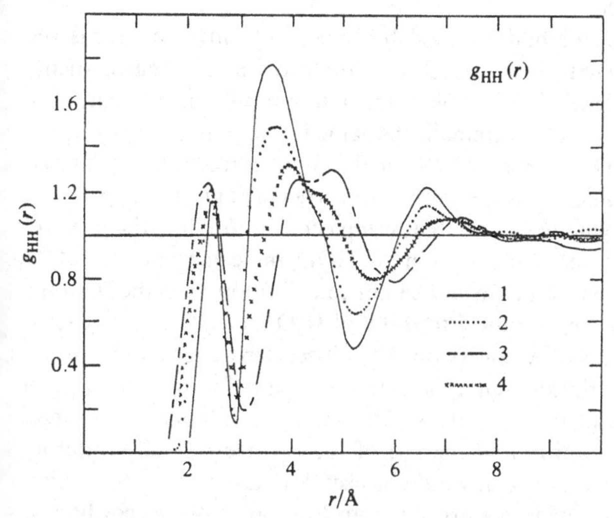

Figure 37. The $g_{HH}(r)$ functions obtained by the UKC group for different combinations of datasets.

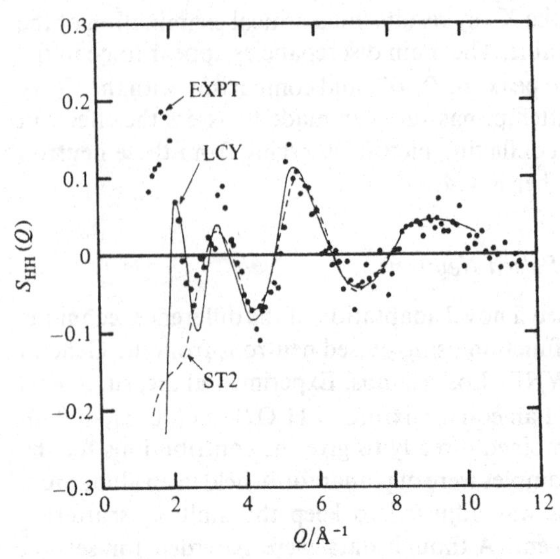

Figure 38. The partial structure factor, $S_{HH}(Q)$ obtained at WNR. [63]

data differ from both sets of results obtained with the reactor technique and the curve is in some respects intermediate between these extremes. The initial peak in $g_{HH}(r)$ is smaller and broader than the other data, although the position is in good agreement; the minimum after the peak is shallower but there is some oscillatory structure in the 4–6 Å region. In a later paper by Soper [64] a comparison is made with the reactor data for D_2O, which reveals significant differences in both the inelasticity corrections and the Q-scale resolution (see figure 8).

The principle of the method adopted by Soper and Silver concerns the reduction of errors due to inelasticity effects by direct difference measurements on the cross-section values. This involves the assumption that the scattering features of both H and D atoms remain the same in the pure liquids and the H_2O/D_2O mixture. The nature of the inelasticity corrections in pulsed neutron diffraction studies is quite complex and is dependent on the spectrum shape as well as the dynamics of the scattering centres. [61] For the mixture there will be a 50 per cent proportion of HOD molecules due to H/D exchange and it is not clear that the incoherent contribution from the H atoms in this combination will be identical to those of H_2O molecules. For reactor measurements the inelasticity effects on the self-scattering terms produce a smooth monotonic variation but this is not the case for pulsed neutron measurements, except at high Q-values. Although the data show no direct evidence of this contribution and the use of the relatively small scattering angle of 40° minimizes this effect it would be useful to have more information on the likely magnitude of terms arising from this variation. According to quoted calculations based on the full theoretical treatment of the inelasticity

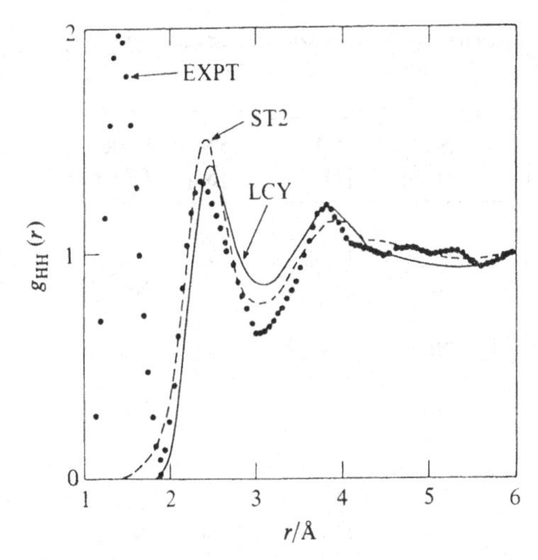

Figure 39. The $g_{HH}(r)$ curve obtained from figure 38.

for a freely translating and rotating molecule, the intramolecular interference terms produce a spurious peak at 0.7 Å. The $g_{HH}(r)$ data obtained from the transform of the experimental data show this artefact, but the magnitude differs from that of the calculation, suggesting that there could be unforeseen problems in the handling of the inelasticity corrections. This factor emphasizes the difficulties of making appropriate corrections in the pulsed neutron scattering studies of samples containing hydrogenous materials. This general technique is clearly of great importance, but further work is needed to clarify fully the procedures for the treatment of the observations under these extreme conditions. In addition, it has long been recognized that the most serious limitation in a pulsed neutron technique based on current instruments is the lack of resolution at low and intermediate Q-values (section 3.2). If the resolution broadening primarily affects the peaks at low Q-values this will produce a general reduction in the oscillatory magnitude of $[g_{HH}(r)-1]$ which will particularly affect the shape at the longer r-values. It is noticeable that the first peak in $g_{HH}(r)$ from the Los Alamos results is comparable in size to that of the UKC data but lower than that of the Oak Ridge data, in spite of a high truncation point, Q_M, of 10.8 Å^{-1}. However, the dip at 3 Å is comparable with the Oak Ridge results and smaller than that of the UKC data, so that there is incomplete agreement with either of the other data sets. The large oscillations in the 5–8 Å region are notably absent, but it is unclear whether this could be explained by the resolution effects in the low-Q region.

6.4. *Comparison of the Partial Functions*

The previous sections have outlined the basic experimental procedures that

Table 7. *Checklist comparison of experimental data for $g_{OH}(r)$ and $g_{HH}(r)$ curves*

Authors... Reference...	PK [**111**]	NT [**119**]	RD [**115**]	SS [**63**]	Sim§ (Various)
$g_{OH}(r)$					
(i) Large peak at ~ 2 Å	No	Yes	Unlikely	—	No
(ii) Deep minimum at ~ 2.3 Å	(Yes)†	No	(Yes)†	No	No
(iii) Peak displacement to ~ 2.9 Å	(Yes)†	No	(Yes)†	—	No
(iv) Oscillatory structure 4–7 Å	Yes	No	Yes	—	No
$g_{HH}(r)$					
(i) Sharp peak at ~ 2.4 Å	Yes	Yes	Yes	Yes	Yes
(ii) Deep minimum at ~ 2.9 Å	(Yes)†	No	(Yes)†	No	No
(iii) Broad second peak (> 1.4 Å)	Yes	No	Variable	No	No
(iv) H and D equivalence	—	Yes	No	—	—
(v) Oscillatory structure 4–7 Å	Yes	No	Yes	No	No

§ The column marked Sim gives a general representation of the features obtained from computer simulation results; different routines show some variations.
† These features are subject to re-interpretation (see section 6.5).

have been used to extract information on the partial pair correlation functions for water. Unfortunately, there is no clearcut picture that emerges from these studies, and a definitive set of functions remains to be established. It is natural that theoreticians should claim greater unanimity in the computer predictions when viewing the apparently diverse results of the experimentalists, but there are also a number of common features in the experimental results which differ sharply from the simulation results. Table 7 gives a convenient checklist of the main features in the experimental curves and a comparison with the overall results of the various simulations. Egelstaff and Root [**120**] have argued that it is inappropriate to compare 'real' water with 'computer' water at the same temperature. By considering the similar shapes of the isochoric temperature-derivative functions they suggest that non-additivity effects can be incorporated by scaling the interaction potential. Other features of the $g(r)$ functions are also considered in this context but the concluding remarks state tentatively that 'it is probable that to achieve overall agreement non-additive terms and quantum corrections will be needed'.

The most obvious paradox to emerge from the comparison of experimental data with computer simulations concerns the local configuration of the molecules. It is already known from the X-ray results that the computer predictions lead to too much tetrahedral ordering of the oxygen positions. In contrast to this behaviour it now seems the neutron results indicate that the simulations underestimate the short-range order in the $g_{OH}(r)$ and $g_{HH}(r)$ partial functions. This seems to imply that the orientational ordering is much stronger than expected from the use of interaction potentials currently adopted, and this may be an indication of the importance of ring structures in stabilizing the orientational correlation effects through collective interactions. The most controversial area concerns the 5–8 Å region. The similarity in the results of Palinkas *et al.* [**111**] and Reed and Dore [**115**] is very striking, particularly as the neutron data were extracted from an over-determined data base. The author must admit to some surprise in this agreement and also a slight bias in favour of these determinations, partly as a result of personal involvement and also for the technical reasons that have been discussed in the earlier sections. However, a consideration of the geometrical features arising from these results leads to some conceptual difficulties; these problems are examined in the next section.

The conjecture that hydrogen and deuterium substitution does not result in isomorphous replacement is not new. There is a large body of literature which points to differences in the physico-chemical properties and spectroscopy of H_2O and D_2O liquids. Stanley and Teixeira [**95**] show that there is an approximate shift of +6 °C in changing from light to heavy water, citing properties such as melting point (0 → 4 °C) and temperature of maximum density (4 → 11 °C) along with other temperature-dependent properties. The relation becomes even clearer when the temperature dependence is extended into the under-cooled region as demonstrated by Angell [**104**] and Lang and Lüdemann. [**105**] As it has already been shown that the 'fragile' structure of water is sensitive to small changes in temperature, it is not surprising that there are some differences due to isotopic effects. The origin of this variation is less easily pinpointed; a classical treatment of the molecular trajectories would presumably give equivalent results for the spatial correlations in both H_2O and D_2O if the interaction potential was assumed to be identical. The differences must therefore indicate that the interaction is affected in some way by factors arising from the change in atomic mass. This latter possibility can arise if there are quantum effects present, and the theoretical aspects of this phenomenon have become a subject of topical interest [**93**, **94**]. Present indications suggest that the changes produced in the computed $g(r)$ functions are insufficient to eliminate the discrepancy with the experimental $g(r)$ data, but this is an area which is attracting much interest and may lead to different conclusions in a few years time. The full consequences have yet to be evaluated, but there now seems little disagreement with the view that a quantum representation will be required if the properties of water are to be properly described.

In terms of the spatial correlation functions, the main effect in changing from a classical to a quantum description will be a broadening of the peaks due to zero-point motion. It is possible that this is the main cause for the inconsistency in the analysis of the $g_{HH}(r)$ correlation functions reported by the UKC group [**115**, **121**] and shown in figure 36, although an alternative possibility is considered later in this section. The replacement of deuterium by hydrogen causes an increased broadening of the atomic location due to the change in mass, so that the principle of isomorphous substitution cannot be fully established. It is difficult to test this conjecture in the case of water, but some interesting neutron diffraction results have recently been reported by Montague, Dore and Cummings [**122**] for H/D isotopic substitution in the hydroxyl group of methanol. The experiments were conducted in a similar manner to those on water and four data sets were obtained for CD_3OD, CD_3OH and two CD_3OH/D mixtures. The analysis revealed similar systematic variations to those for water and suggested that the hydrogen atom had a broader distribution than the equivalent deuterium atom in the hydroxyl position. It would be premature to deduce that this effect is prevalent in all hydrogen-bonded systems, but it is clearly of great importance to make further studies on other suitable liquids. The problems with the treatment of water data arise from the fact that each molecule has two exchangeable H/D positions, so that the intramolecular and intermolecular correlation functions are affected in a very complex manner. The current analysis [**9**, **110**] of some recent neutron experiments on liquid DF and an HF/DF mixture seems to provide an even more striking picture of this behaviour. Hydrogen fluoride exhibits one of the most strongly associated interactions through hydrogen bonding. The results appear to provide evidence for ring structures with well-defined hydrogen-bond distances, but the peak heights for H and D correlations are not in the correct proportions, and this seems to be related to some anomalous features seen in an electron diffraction study of HF vapour. [**101**] Delocalization along the bond [**123**] becomes a distinct possibility through the formation of a resonance hybrid. Further studies of this phenomenon are obviously required and could have a bearing on the interpretation of the water results.

6.6. *Geometric Relations in the Hydrogen-Bonded Neighbours*

The observation of well-defined peaks in the various partial functions of liquid water or the composite pair correlation functions for amorphous ice provides useful information for deducing the spatial and orientational relationships between two hydrogen-bonded molecules. It is possible to construct a model which is representative of the mean atomic locations as given by the peak positions and to evaluate the associated angles between the various bond directions. Figure 40(a) shows a schematic diagram involving the tilt angle ϕ in the plane of a hydrogen-bonded neighbour. The various distances marked on the diagram are 'average' values for the liquid phase. It is immediately apparent that $r_{O_1O_2} \sim r_{O_1D_1} + r_{D_1O_2}$, which indicates that the hydrogen bond

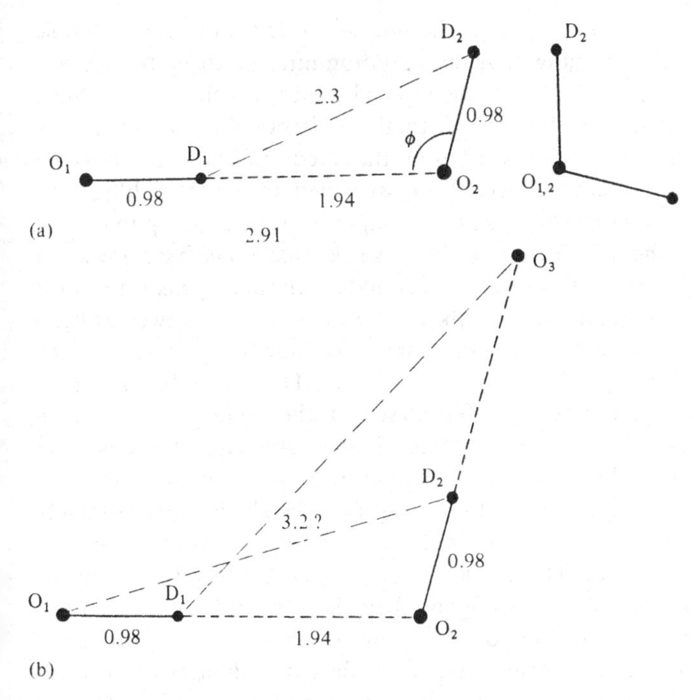

Figure 40. Geometrical effects in the hydrogen-bonded configuration for two molecules; all distances are in Å.

itself shows very little deviation from linearity. Using the value for $r_{D_1D_2}$, which is the first peak in the $g_{DD}(r)$ function gives a value of 98.3° for the tilt angle ϕ. This value is significantly lower than the tetrahedral angle of 109.5° and the intramolecular DOD angle of 104°. This behaviour suggests that there is a closing of the ring structures which is probably responsible for the increased density of the liquid over the solid phases. The reduced angle is close in value to the angle of 97° given by Hajdu [**101**] in the asymmetric distortion of tetrahedral symmetry in the tetragonal form, and further investigation of this geometry would be instructive.

The network structure of figure 40(a) can be further extended as shown in figure 40(b) to include additional atoms of the same or near-neighbour molecules. The broad $g_{OO}(r)$ peak obtained from the X-ray results is consistent with a spread of angles, and the temperature variation studies suggest that the $r_{O_1O_3}$ value increases as the temperature decreases. This feature conveniently illustrates the way in which the network opens out to give the lower density in the under-cooled liquid and also the approach to a tetrahedral value seen in the data for amorphous ice. The structure shown in figure 39(b) defines the $r_{O_1D_2}$ value at 3.2 Å. Both data sets from Palinkas *et al.* [**111**] and Reed and Dore [**115**] exhibit a sharp peak at a much shorter value of 2.9–3.0 Å. It is difficult to see how this value can possibly be achieved without moving the position of the O_1 atom away from the line of the D_1O_2 direction. Simple geometrical calculations show that the deviation introduced by this change would have too large an effect on the O_1O_2 distance, which is well defined by the x-ray data. It therefore seems that it is impossible to construct a network which simultaneously satisfies all these requirements. The possible introduction of 'shrinkage effects' **[124]** as noted in electron diffraction studies of molecules with large librational amplitude does not help to resolve the discrepancy. The only feature in the network which does have a peak at this position is $g_{OO}(r)$ and this value corresponds to the main peak for nearest neighbours! A close examination of the $g_{OD}(r)$ curves for these two data sets indicates that there appears to be a superimposed contribution which has the sharp features of the $g_{OO}(r)$ curve at low-r values. The reason for this effect can be seen by comparing the $g(r)$ curve for the neutron results with $\langle b_{HD} \rangle = 0$ (figure 35) and that for the X-ray measurements. There is good agreement over the full r-value range except at the position of the main peak, where the neutron results underestimate the peak height by approximately 35 per cent, probably due to resolution effects from the low cut-off point, Q_m, chosen for the transform. This reduction in peak height therefore seems to cause a systematic error in this region when the separation of the partial functions is carried out, leading to the appearance of a spurious peak in the $g_{OH}(r)$ curves. The $g_{HH}(r)$ results are also affected and the oxygen peak features are seen in an inverted form giving a deep minimum at 2.9 Å. Having recognized the importance of the oxygen contribution to the analysis it is instructive to consider the assumptions made in the treatment of the data. Emphasis has been made of the differences resulting from H and D variation without considering whether this isotopic change could also influence the $g_{OO}(r)$ curve. Examination of the separate curves for large r-values indicates that there may be systematic effects which relate to both the $g_{OO}(r)$ and $\Delta g_{OO}(r, T)$ curves; the latter contribution can be expected from the effective temperature shift, which is correlated with the isotopic change and may affect the behaviour at larger r-values. However, the contributions are likely to be small in the $g_{OH}(r)$ curves, and although they may reduce the oscillation amplitude of the 5 Å positive peak they will not remove the oscillatory structure in this region.

The $g_{HH}(r)$ curves can be scrutinized in a similar manner. There is much more variation in these curves for the high r-value region. This is to be expected because of the greater variation in the H/D proportions (table 6) which apply to the different cases for separation process. The fact that the $g_{OH}(r)$ data are in excellent agreement with each other and the $g_{HH}(r)$ data are not is, itself, a very strong indication that isotopic differences are present. The $g_{HH}(r)$ results show differences which relate in a systematic way to the H/D proportions. The variation in the shapes of the curves above 5 Å resembles that of $\Delta g_{OO}(r, T)$ in this region. If it is assumed that the change from H to D is equivalent to a temperature shift, it is possible to see why these curves show some variation, although the effect looks much larger than would be expected. It is also noticeable that there are systematic variations in the initial rise of the first peak in $g_{HH}(r)$ which correlate with the H/D proportions and suggest that the DD peak is possible at a slightly shorter distance than the HH peak. This is at first sight a surprising observation, but could arise quite naturally from the isotopic difference effect resulting from spatial delocalization along the hydrogen bond. Clearly, these trends require a very careful assessment in relation to both the random and systematic errors which are inevitably present in the data treatment, but the apparent internal discrepancies in the UKC data can be rationalized within the context of accepted principles. The predicted isotopic differences for the oxygen positions are obviously susceptible to further study by X-ray measurements and it is quite possible that these subtle effects have gone undetected in earlier work. The precise pressure studies of Neilson **[57]** provide a clear pointer that H_2O and D_2O should not be regarded as structurally identical.

The results of Palinkas *et al.* **[111]** show similar effects which cannot be explained by this means, since D_2O was used in all three experiments. The sharp peak at 3.0 Å possibly arises from the temperature variation in these studies, since the D_2O data used in the analysis were taken at 25 °C. The temperature variation studies of Bosio *et al.* **[11]** using X-ray diffraction indicate a differential shift in the peak position of $g_{OO}(r)$ which could influence this region, but the effect again looks too small. Another possibility concerns the use of the form-factors for the description of the molecular electron-density distribution, since one presumes that the strong hydrogen bonding revealed by these experiments must have some important effects on the shape of the form-factors. However, they are unlikely to introduce features with a sharp spatial distribution, and it is difficult to see how they could be quantitatively evaluated to the required accuracy. The surprising agreement in the results of these two independent experiments raises a number of problems for the geometrical interpretation. Both data sets appear to show correlations with the spatial distribution of oxygen positions, but these are insufficient to remove the effects completely. The oscillatory structure at large r-values does seem to relate well to the expected variation of second-neighbour orientational correlation, which occurs with change in temperature and evolves towards the amorphous ice structure. It would seem that the local tetrahedral ordering of adjacent molecules undergoes fairly minor variations with the imposed conditions. It is the way in which these small changes are amplified in the second nearest hydrogen-bonded molecules that has the most direct influence on the overall properties.

7. Conclusions and Future Projections

It is approximately fifty years since the first conjecture was made concerning the structural characteristics of liquid water. The pioneering work of Bernal and Fowler **[125]** in 1933 laid the foundations for a research topic that has continued to mature over the intervening years in a steady and progressive manner. It might sometimes be claimed that our state of knowledge has made only limited progress since that foundation paper and that the basic principles presented over fifty years ago remain 'in vogue' just as when they were originally formulated. It is patently obvious that there have not been the epoch-making breakthroughs that punctuate the conceptual development and formulation of new pathways in the more expansive research fields such as elementary particle physics, molecular biology and radioastronomy. Nevertheless there has been a gradual evolution in the understanding of this complex and commonplace liquid which we simply call 'water'.

The quantitative work on the structural behaviour described in the previous sections has emphasized some of the unique properties exhibited by water while simultaneously suggesting that these characteristics result quite naturally from the specific features of the molecular interactions which create and possibly distort the basic tetrahedral nature of hydrogen-bonded configurations. Water therefore emerges as an unusual liquid, but one that should be satisfactorily described by the normal procedures of chemical physics incorporating a theoretical formalism which would be equally applicable to any other molecular liquid. The limits to our understanding appear to rest entirely within the realm of the hydrogen-bonding interaction, and the experimental results presently available suggest that future progress will require a much more precise investigation of this phenomenon.

On a microscopic scale it is evident that an assembly of water molecules possesses a very specific local environment in which adjacent molecules adopt preferential positions. The strongly correlated orientational behaviour revealed in the structural properties of amorphous ice and under-cooled water, coupled with the ideas of a description based on percolation concepts, indicate that the collective behaviour of the molecular assembly is predominant in defining the properties. The existence of transient water clusters has in some respects reconciled the apparent conflict between the continuum and two-state models presented in the earlier attempts to give a descriptive formulation of the properties. The conceptual principles involved in this approach will presumably undergo further development, but there is clearly a problem in relating a formalism which emphasizes network structures and connectivity in a fictitious 'hydrogen-bond space' to the spatial distribution functions of the real liquid. There seems little doubt that the anomalous properties of under-cooled water arise from the network-building capacity of the tetrahedral bonding, but this conceptual framework remains very restrictive in its ability to make quantitative predictions in relation to the spatial correlations in real water. The fact that the fundamental predictions are insensitive to the formal description of the hydrogen bond is itself an indication that the theory is of an essentially descriptive rather than prescriptive nature and may be unable to provide detailed information at the molecular level. The effect of temperature may simply alter the value of p_B in the percolation model, but in real space there is no clear distinction between bonded and non-bonded molecules and p_B is itself an average property of the ensemble. Some of the computer graphics films showing the librational motion of water molecules have indicated how the hydrogen bonds appear to switch on and off according to whether the characteristics of the local environment satisfy the chosen criteria. This phenomenon was at one stage thought to give some support for the 'flickering cluster' model, but the real point of the video demonstration is to show that the choice of a satisfactory hydrogen-bond criterion is intrinsically difficult and eventually must be based on arbitrary principles.

The reality of the situation is that an individual molecule is in a changing environment in which the 'degree of hydrogen bonding' is continuously variable. In terms of the structural configuration it is probably not the immediate neighbours which define this property but the establishment of ring connectivity in which all molecules are at appropriate orientations for hydrogen bonding to be apparent. As the molecules translate and rotate/librate there will be variations in the relative orientations, creating unfavourable angles for the bonding which eventually result in some member(s) of the ring adopting orientational positions in which the hydrogen bonding to alternative molecules becomes more favourable. Whether this occurs 'suddenly', as in the jump model, is obviously a matter for further investigation by spectroscopic means. The 'hydrogen bonding' of any one molecule may therefore be viewed as a continuous time-dependent variable in which each molecule participates in an exchange of 'connectivity' with adjacent molecules. This descriptive phenomenon is obviously linked more with the temporal evolution of hydrogen/deuterium atom positions, as described by the dynamic structure factor $S(Q, \omega)$ rather than the static structure factors $S(Q)$ which have been considered in this review. Spectroscopic studies have already given much valuable information on hydrogen bonding, and some of the new studies of temperature-dependent effects in quasi-elastic neutron scattering will inevitably add further details over the next decade. There is much to be gained by the combination of experimental data on structure (elastic scattering) and dynamics (quasi-elastic and inelastic scattering). The satisfactory integration of these diverse techniques is probably the main task for the next decade, but this use of complementary information in the building of a more comprehensive picture will not be easily achieved.

The present position on the characterization of the microscopic structures remains unclear at present, although a possible resolution of the discrepancies in the experimental data on the partial $g(r)$ functions is envisaged and will require further calculations for verification. The experimental work on D_2O has provided valuable information on the structural changes due to temperature and pressure variation, but is relatively insensitive to the important region of 1–3 Å due to the compensating oscillations (figures 34 and 37) of the OD and DD contributions. As a result, the spatial characteristics of the hydrogen bonding to neighbouring units are partly concealed in the composite

$d_L(r)$ function for neutron studies of D_2O. It is therefore imperative that further precise information is obtained on the partial correlation functions for this region. Although there are discrepancies in the results of different experiments, there is general agreement in the main features at low r-values (table 7 and section 6.5), and data sets emphasize that the hydrogen-bonded OD distance is separated from the rest of the curve by a deep minimum. The experimental results are consistent in suggesting that the local hydrogen-bonded configurations are much more structurally ordered than the simulations suggest.

The network structure of amorphous ice, the temperature-dependent variations of the liquid structure and the transient clusters revealed in the SAXS experiments, all carry the suggestion that an effective two-body interaction potential will finally be inadequate in providing detailed agreement with the observations. This has been known for some time, but it is not easy to see how the next step can be taken to remedy these theoretical limitations. Nevertheless, it must be stated firmly that no computer simulation seems able to give a completely satisfactory representation of the three-pair correlation functions or their variation with temperature and pressure. This situation contrasts strongly with the description of spatial correlation functions for other molecular liquids which do not exhibit hydrogen-bonding effects.

The isotopic variations also require further specific investigation. The UKC analysis suggests that hydrogen and deuterium are not equivalent and that the interpretation of the H/D substitution experiments is not straightforward. This viewpoint carries an implicit indication that quantum effects play a role in modifying the behaviour. All groups have concentrated on mixtures with a relatively large hydrogen content in order to exploit the wide variation in the $\langle b_{HD} \rangle$ values. The large corrections which apply to the hydrogenous samples contribute to the systematic errors and appear to lead to significant perturbations in the spatial correlations through the isotopic difference effect. It is possible to conduct similar experiments to separate the partial functions using smaller proportions of hydrogen provided that sufficient statistical accuracy is achieved; the smaller contribution from incoherent scattering also helps to make these experiments feasible, but no attempt has yet been made to obtain this information. The study of isotopic effects is not restricted to measurements on water alone, and the work on methanol mixtures has probably set the scene for a series of studies on systems in which single hydrogen-bond interactions are predominant. This situation is much more favourable for detailed study because the interchange of hydrogen and deuterium in the hydroxyl group has a smaller effect on the overall properties of the mixture but still permits the use of both single- and double-difference techniques to isolate some of the partial pair correlation functions.

It is apparently the space-filling property of the double-donor, double-acceptor feature of the water molecule which gives it particular properties, such as a maximum in the temperature dependence of the density. **[113]** There is therefore a conjunction of two main properties which appear to be required to give water its characteristic behaviour, and these are the fourfold tetrahedral nature of the association with neighbouring atoms and the ability to create a strong correlation of orientation in the local environment through hydrogen bonding. The various structures of the crystalline phases of ice are a convenient reminder that the pattern of hydrogen bonds can adopt many stable configurations which satisfy these criteria. Recent studies [**87**, **88**], however, have shown that the positions of protons along the bond axis can exhibit significantly asymmetric profiles. In the higher-temperature region of liquid water it seems inevitable that proton transfer across the mid-point of the bond becomes more probable, although the increased disorder and eventual disruption of the network would tend to inhibit this process at very high temperatures. The simultaneous transfer of all six appropriately placed protons in a cyclic hexamer (figure 41(a)) would not cause any change in the structural characteristics as defined in terms of the time-averaged positions seen in a diffraction study. The only possible indication of this dynamic process would be through an effective asymmetric broadening of the OD peak, such as shown in figure 14(b). Current experimental methods are unable to probe this property with sufficient precision due to the lack of spatial resolution and uncertainties arising from the inelasticity corrections, but existing data suggest that this phenomenon could occur. The available evidence indicates that the peaks are broader than would be expected from the classical picture for normal mode vibration of the molecule. It is already known (section 4.1, table 3) that the mean r_{OD} bond-length varies systematically with the state of the material, and this could readily be interpreted as an asymmetric distortion of the probability profile, in much the same manner as observed in ice Ih. If the classical picture of point-like particles is replaced by a quantum one in which the atoms are characterized by a wave function with a broad spatial distribution it is not unrealistic to consider that the proton could be delocalized along the hydrogen bond. In ice and water the presence of many interconnected ring structures seems to show a similar effect, and conveniently incorporates the collective nature of the interaction which seems to be an important feature of the behaviour. In this context the properties of under-cooled water seem more appropriately described as those of a 'quantum gel' than a molecular liquid. In the low-temperature regime there are already several pointers that gelation features may be apparent, and other characteristics emphasize the inadequacy of the classical treatment.

If water does require a quantum description based on delocalization of atom positions along the hydrogen bonds through cyclic rings there are a number of consequences which are open to further study in both water and ice structures. Resonant transfer around the ring, presumably, is not only dependent on the correct positions of all the atoms but also assumes that they are of similar mass. If H/D mixtures are used cyclic rings will contain 'impurity' atoms which will interfere with symmetry and presumably affect

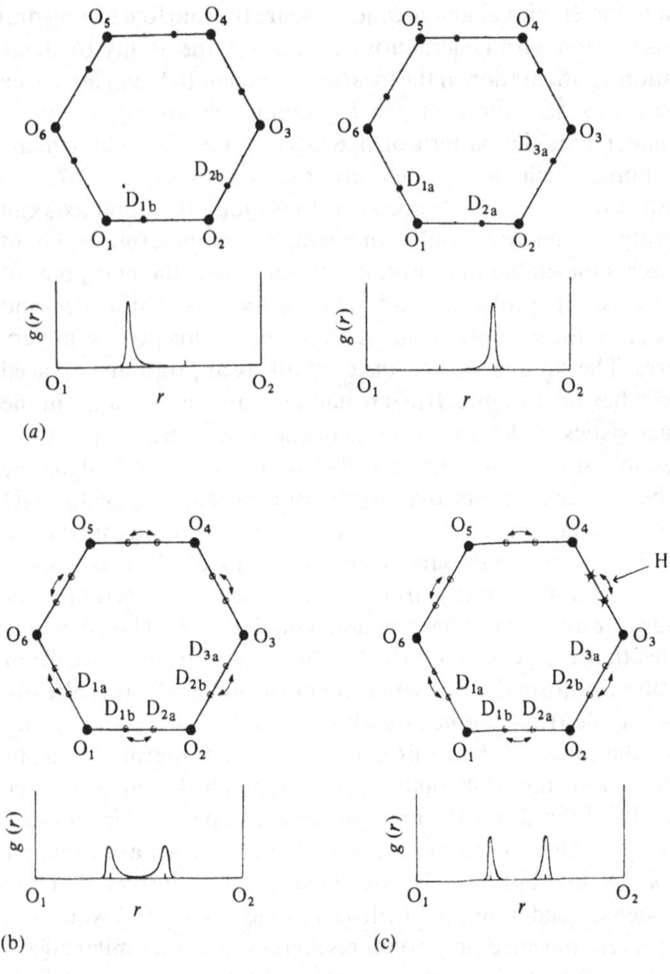

Figure 41. Ring structures from hydrogen-bonded interactions: (a) classical picture of cooperation motion; (b) quantum picture of proton delocalization; (c) effect of 'impurity' atoms or bond defects.

both exchange properties and spatial distribution functions. Under these circumstances a small proportion (~ 10 per cent) of H_2O in D_2O will have a significant effect, and the diffraction pattern should show a non-linear behaviour with respect to α_D. It is even possible that this phenomenon has been inadvertently observed in some of the early neutron studies of liquid D_2O. Samples available in the early 1970s usually contained up to 2 per cent of H_2O isotopic impurity. The diffraction measurements frequently showed a weak shoulder on the high Q-value side of the main diffraction peak which was not readily reproducible in different runs. Since this variation made a negligible contribution to the $d(r)$ or $g(r)$ functions it was not regarded as important, and with the advent of samples with higher isotopic enrichment the difficulty apparently disappeared. In retrospect this neglected information may have represented a small but real effect, particularly when one notes that this region of the curve corresponds to the second peak for $\langle b_{HD} \rangle = 0$ and is also revealed as a very sensitive region of the $D_M(Q)$ curve extracted from neutron measurements of water in different porous media. [**26**]

The picture which emerges from this kind of speculation raises complex questions, and it is possible that the study of low-temperature water has produced, and will continue to produce, information which is difficult to interpret in detail. Water is expected to behave like a normal liquid at high temperatures when the hydrogen bonds are substantially broken. There is a school of thought which suggests that a more logical approach to the problem is to study the liquid in a regime where the properties are more equivalent to those of other molecular liquids and can be understood in conventional terms. Unfortunately, this means a liquid water sample under superheated conditions where the corrosive properties make containment extremely difficult. In principle, it should be possible to define the pair interaction potential from measurements under these conditions so that the importance of the 'additional terms' which seem to be operative at lower temperatures could be more clearly evaluated. Very few attempts have been made to follow this path in spite of the wide-ranging X-ray studies of Narten and Levy in the early 1970s, but it remains a challenging prospect for study and it is possible that the fixed angle counters used in pulsed neutron methods offer some technical advantages over corresponding measurements using reactor sources. Egelstaff [**120**] has stressed the importance of temperature and pressure in probing the systematic behaviour, but only limited experimental data are available for the variation of both parameters simultaneously. Experimental measurements at constant density over a wide region of the P–T diagram would be very instructive.

This review has attempted to show how diffraction methods have been used to investigate the 'structure' of water. It sometimes seems that the more we learn from our experimental studies, the less we understand about the microscopic properties of this important material. It still presents to us a fascinating enigma in which we have only just begun to realize the immensity of the problem we are tackling. Progress is inevitably slow and painstakingly methodical, but the challenge can be met and will no doubt be reported in further volumes of this journal. There are probably some relatively simple answers to be obtained in the 'quantum gel' regime of low-temperature water or the discrete particle nature of high-temperature water. It is just one of those infuriating facts of science that our main interest is most firmly fixed on that intermediate regime of 'normal' water which, by accident or design, so clearly influences our lives.

Acknowledgements

It is a pleasure to acknowledge the many friends and colleagues who have contributed in various ways to this review. Within the neutron scattering group at UKC, three research students at different times have wrestled with technical and computational aspects of the work before moving on to other fields; they are Drs Jeff Walford, Ian Gibson and Jim Reed. Others supported by SERC funds who have played a role in the developments are Drs John Clarke, Mizan Chowdhury, Jack Wenzel and David Steytler. I would also like to thank Dr Ian Page (AERE) for introducing me to the world of neutron scattering and Professor Jack Powles (UKC) for a continuous flow of advice and information on the theoretical aspects of the inelasticity corrections. Amongst colleagues elsewhere, who are too numerous to number, I would particularly wish to thank Drs Pierre Chieux (ILL), George Neilson (Bristol) and José Teixeira (CEN, Saclay) for invaluable participation in colloborative work. Finally, this evolving programme of research would not have been possible without the financial support of the Neutron Beams Research Committee of the Science and Engineering Council for an extensive period up to 1983.

My thanks are also expressed to Betty Jones who has painstakingly typed the initial draft and carried it through its various stages to emerge as a final manuscript.

Postscript

Since this manuscript was prepared there have been a number of developments concerning the use of neutron scattering for studies of water and ice.

(a) Recent quasi-elastic neutron scattering measurements of Teixeira, Bellisent-Funel, Chen and Dianoux [**129**] on H_2O have provided further information on the hydrogen-bond dynamics as a function of temperature. The prediction of a collective mode in the form of hypersound obtained from MD simulations has very recently been confirmed by Teixeira and Bellisent-Funel [**130**]. A review of current spectroscopic information for normal and super-cooled water has been prepared by Chen and Teixeira [**131**]. The recent work seems consistent with percolation concepts (section 5.3).

(b) Analysis of neutron diffraction data [**132**] taken on a D_2O emulsion sample at -29 °C has given strong evidence for the continuing evolution of the disordered tetrahedral network at low temperatures. The displacement of the diffraction peak is in satisfactory agreement with the tentative extrapolation shown in figure 28; a preliminary description of these data has been given by Dore [**133**].

Experimental work on H_2O/D_2O mixtures has been further developed along the lines indicated in section 6; diffraction measurements have now been made for $\alpha_H = 0.03$, 0.05, 0.10 and 0.20 at 20 °C in order to extract a set of

$g_{OD}(r)$ and $g_{DD}(r)$ where the problems of representing the scattering by hydrogen have been minimized.

(c) An analysis of high-precision neutron data for ice-Ih by Kuhs [**134**] has revealed the presence of oxygen disorder. The results show that the true equilibrium value (r_e) for the intramolecular bond is 0.98(5) Å, which is much less than that quoted in table 3 and in good agreement with that for other condensed phases.

Appendix 1: Notation Used in Various Descriptions

There are unfortunately two descriptive systems used in the theoretical formalism for the definition of structure factor and its related components. The system used in the text is based on the use of $S_M(Q)$ to define the molecular structure according to a notation originally introduced by Egelstaff [**129**] and also used by Powles [**3**]. In the alternative form used by Narten [**19**] for both neutron and X-ray diffraction the scattering vector is represented by the symbol, k, and the total structure factor as $S(k)$. In each case the structure factor is divided into separate parts resulting from self-terms ($i = j$), intra-molecular terms ($i \neq j$; same molecule) and inter-molecular terms ($i \neq j$; different molecule). In the alternative notation, the terms where $i \neq j$ are referred to as 'distinct' contributions to distinguish them from the self-terms and correspond to the interference contribution referred to in the text. The correspondence between the different symbols is given in table A1. The subscripts α, β used to describe different species of atom are used consistently in the two systems. The representation in real space utilizes a different form in which the new symbol, $h(r) = g(r) - 1$ is introduced. The expression which is given as an equation:

$$d(r) = 4\pi r\rho[g(r) - 1] = \frac{2}{\pi}\int_0^\infty Q[S(Q) - 1] \sin Qr \, dQ$$

Table A1.

Symbol (text) [E & P]	Symbol (equivalent) [B & N]
Q	k
$S(Q)$	$S(k)$
$D_M(Q)$	$\hat{h}(k)$
$g(r) - 1$	$h(r)$

The equivalence of symbols used in the text following the notation of Egelstaff and Powles [E & P] with that used in the later papers of Blum and Narten [B & N].

may therefore be re-written in the alternative notation as

$$h(r) = \frac{1}{2\pi^2\rho}\int_0^\infty k^2 h(k) j_0(kr)\mathrm{d}k;$$

here $h(k)$ is the distinct contribution in k-space. These equivalent expressions apply to a composite $g(r)$ function with various components or the individual partial correlation functions themselves, i.e. $g_{\alpha\beta}(r)$, $h_{\alpha\beta}(r)$ and their transformed counterparts $h_{\alpha\beta}(k)$.

Appendix 2: Inelasticity Corrections

The original treatment of inelasticity corrections was made by Placzek [**14**] for monatomic liquids. The extension of the formalism to molecular systems follows similar principles but introduces many complications because the motion of individual scattering centres is correlated. The various groups that have tackled this problem have adopted a free-molecule approximation in which the molecular unit is treated as an individual entity in which the relative movement of the nuclei is described by a normal-modes analysis of vibrational motion at the appropriate temperature. Corrections evaluated for the self-terms ($i = j$) in the cross-section are dependent on the incident neutron energy since this determines the upper cut-off point in the integration over the frequency ω in eqn (2.18). In principle, it should be possible to make exact calculations for any given scattering situation provided the dynamic structure factor $S(Q, \omega)$ is defined and the experimental conditions are specified. In practice, there are differences in the predictions due to the nature of the approximations. Even for the simplest case of diatomic molecules there is controversy over the evaluation as shown by Egelstaff and Soper [**60**], who compare their results with the treatment of Powles [**127**] and Blum and Narten [**128**]. It is inappropriate to discuss here the technical details of this work, and reference should be made to the original papers. Under these circumstances it is not surprising that there are differences in the application of inelasticity corrections for water. All groups also assume that the corrections apply only to intra-molecular terms which affect the $f_1(Q)$ form-factor but do not influence the $D_M(Q)$ function. This appears to be a satisfactory procedure for D_2O but has not been experimentally tested for H_2O or H_2O/D_2O mixtures.

The use of pulsed neutron techniques increases the complications because of dependence on the spectrum shape and geometrical effects which both influence the time-channel of the detected neutrons. Powles [**129**] has given a general treatment for this case and the effect on the self-terms has been demonstrated for diatomic systems by Howells. [**61**] There are no details of calculations for D_2O but Soper [**64**] has presented some information on the systematic effects occurring in the treatment of the experimental measurements. Clearly there is scope for further theoretical and experimental work on this topic.

Appendix 3: Statistical and Systematic Errors in the Data Analysis

There are several operations which must be carried out to convert the raw experimental observations into a suitably normalized cross-section, intermolecular interference function and finally a pair-correlation function. The computations used to extract the required information can vary in complexity, and it is often difficult for authors to give an indication of the uncertainties that may be present in the final results. The associated 'margin of error' can be of crucial significance when comparing the experimental data with a model simulation. This section attempts to give a brief survey of the possible sources of error, the consistency checks that can be applied and the consequent effects on the final data.

All modern diffraction equipment uses a neutron counter of some form, so the basic information is obtained as a total count for a fixed amount of beam defined by the integrated count of a beam monitor. In early experiments discrete detectors were used, but it is now more usual to employ multi-wire counters, position-sensitive detectors or detector arrays to record the data. With these new systems to improve the data collection efficiency it is necessary to make regular calibration checks to determine the relative efficiencies and also to ensure that they do not change with time. With modern electronic techniques, a highly stable system has been evolved, and these instabilities are only important in experiments where small differences are being measured. In the case of pulsed neutron diffraction studies there is an additional requirement to define the intensity profile $I(\lambda)$ of the incident beam, and this is usually done with a standard vanadium sample. Background contributions are usually very small except for low scattering angles, where the counter may be close to the incident beam time. For studies of water, the scattering from the container is quite small and can be easily subtracted except in the case of pressure studies, where special methods must be used (section 4.4). The shape of the diffraction pattern for the sample is primarily dependent on the counting statistics and can normally be defined to an accuracy of better than 0.2 per cent per 0.1 Å^{-1} incremental step-length in relative terms using modern instrumentation on a high-flux reactor. The normalization to an absolute cross-section scale is less accurate due to uncertainties arising from self-attenuation, absorption and multiple-scattering corrections. The multiple-scattering may be computed by Monte Carlo methods and the overall effect on the curve is to increase the amplitude of the oscillations without changing the general shape (figure 42). The corrections can be reliably evaluated for D_2O samples but become less accurate for samples with a high scattering power. Any resulting errors will take the form of increased or reduced oscillation amplitude in the interference contribution.

The cross-section data are fitted at high Q-values by a molecular function which contains several parameter variables. For materials containing light atoms like D and H, the fall-off in the cross-section is very important, and

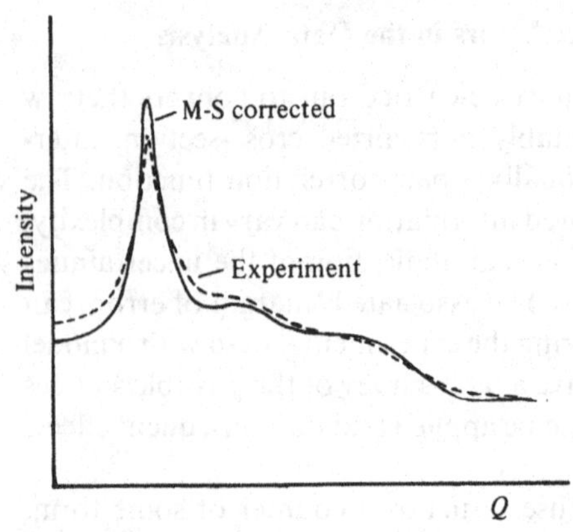

Figure 42.

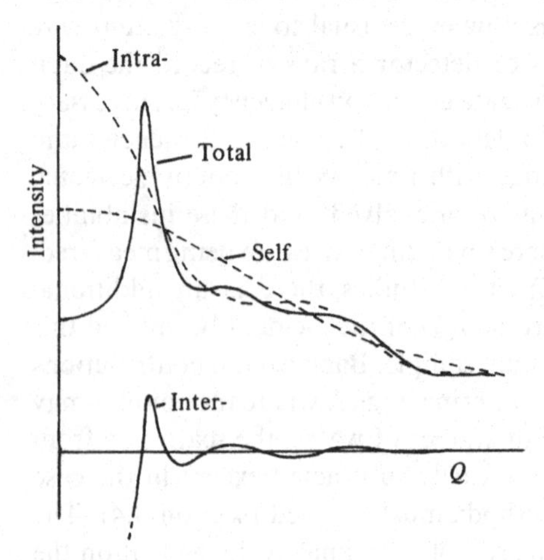

Figure 43.

systematic errors can result in either the self-term description or the molecular interference terms as shown schematically in Figure 43. It can be appreciated that the general form of the pattern is normally retained and the systematic errors mainly affect the amplitude of the $D_M(Q)$ function rather than its shape. Resolution effects are negligible in reactor studies but may sometimes be important for pulsed neutron measurements in the low Q-value range. When the $D_M(Q)$ data is transformed to r-space all sense of the random and systematic errors is apparently lost in the production of the final curve but there are several important check procedures which will indicate certain kinds of systematic error. These are shown schematically in figures 44 to 46 and correspond to incorrect self-term subtraction giving a vertical displacement to $D_M(Q)$, incomplete removal of the intra-molecular contributions producing

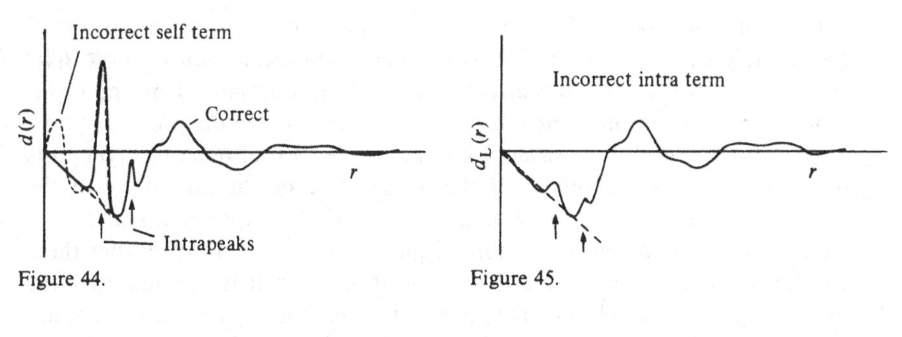

Figure 44. Figure 45.

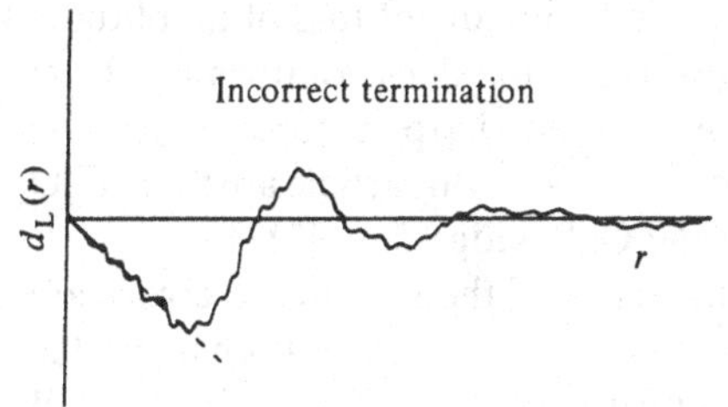

Figure 46.

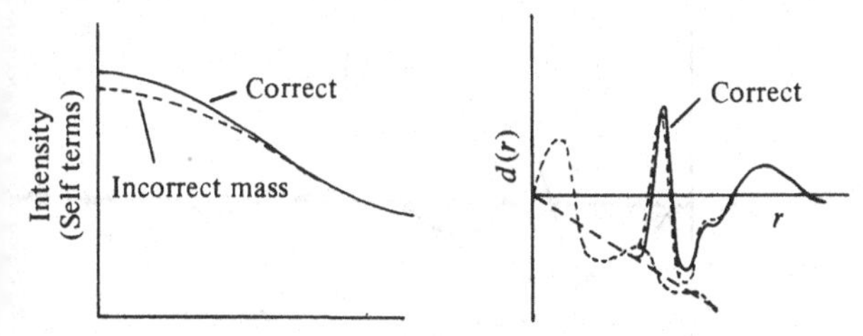

Figure 47.

a sharp residual peak at a known bond-length and inappropriate choice of termination point, Q_m leading to a ripple frequency $Q_m/2\pi$. The low r region is particularly susceptible to small changes in the overall level of the $D_M(Q)$ function and often gives a series of spurious peaks. Since these are well outside the expcted physical range they do not affect the interpretation of the results, but they may indicate that a small adjustment in one of the parameters is needed. In the case of D_2O or H_2O this region provides a good test of whether the inelasticity corrections to the self-terms are of the correct magnitude. A systematic discrepancy is shown in figure 47 and it can be seen that the main result will be a difference curve with a periodicity of approximately $Q_m/2\pi$. Since Q_m is typically greater than 10 Å^{-1} this gives a broad peak at approximately 0.3 Å^{-1} which is immediately recognized as spurious. Some groups used iteration methods which force the curve to follow the correct density line, but this can produce small systematic changes in the overall pattern and must be carefully checked for consistency. The general slope of the line at low r provides a good check on the overall scale factor for the $d_L(r)$ curve, but this curve should already be correctly normalized by the intra-molecular contributions which have already been fitted in the previous stage, and so this is a useful consistency check. Conversion of $d_L(r)$ to $g_L(r)$ plots requires a correct assessment of the density line or the amplitude of the oscillatory structure will be wrongly scaled. If the experiments and data analysis are carried out in the correct manner it is very unlikely that there will be any discrepancies in the peak positions, but it is possible that the overall oscillation amplitudes may differ. A reduction in the truncation point, Q_m, may cause a broadening of the peak and consequent loss of amplitude, but for the inter-molecular terms the $D_M(Q)$ function does not normally have much structure above 6 Å^{-1} so that there are no sharp features in the $d(r)$ curve. The two exceptions to this situation covered in this article are for a-D_2O [**78**] and the sharp peak in $g_{OD}(r)$ from the Oak Ridge data [**119**].

For difference function measurements many of the systematic errors are eliminated to first order, although care must be taken to include changes due to the density of the liquid and the container expansion. The main effect of

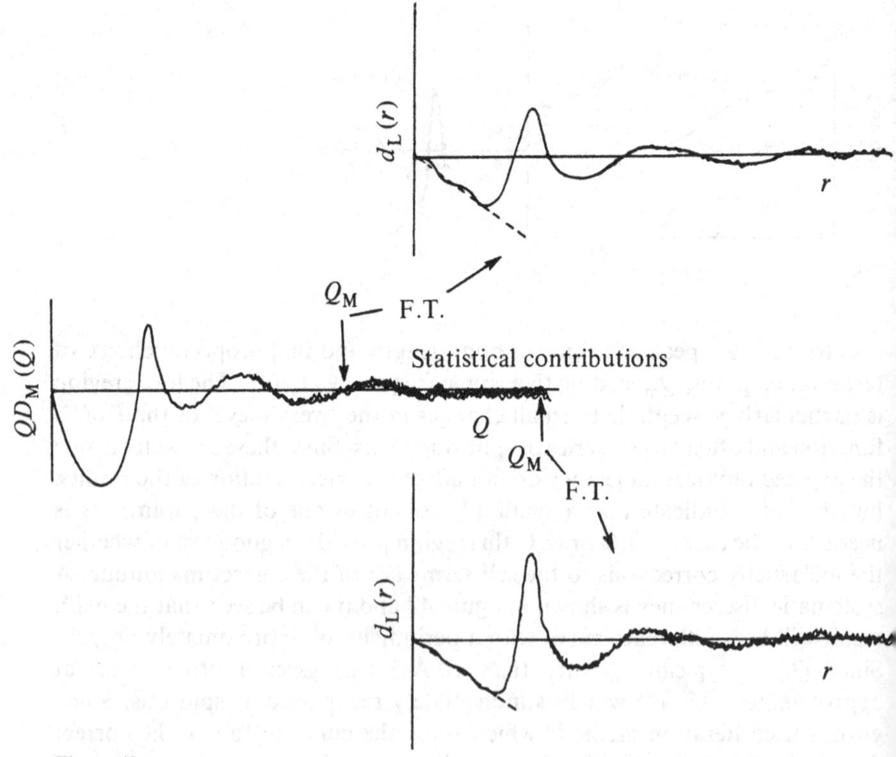

Figure 48.

statistical errors usually occurs in the high Q-value region where a small oscillatory signal becomes lost in the fluctuations arising from the counting statistics. The random count will transform to a noise signal in r space (figure 48). As a result, there is usually an optimum value for Q_m which eliminates unwanted noise at high Q but does not 'throw away' useful information from the signal. It is possible to use smoothing routines, but most groups avoid this procedure as it can introduce bias into the results in a way that is not readily identified.

Finally a brief comment about errors in partial functions is necessary. It is clear that small systematic uncertainties in the different composite data sets can lead to very large discrepancies in the isolated partial functions obtained by solving the simultaneous equations. It seems likely that this is the reason for the widely different conclusions reached by the OR and UKC groups (section 6.4). There is clearly a complex interlinking of systematic errors arising from inelasticity and multiple-scattering effects with the relative statistical accuracy for each of the measurements. When these are combined to give a particular $g_{\alpha\beta}(r)$ curve the final function is highly dependent on various weighting factors and a clear definition of the errors becomes very difficult. The available results should obviously be in better agreement than is apparent at the present time, and it is to be hoped that a careful comparative analysis of the information will help to resolve the problem. Alternatively, there may be further experimental work which will eventually make these initial measurements superfluous.

References

1. F. Franks (ed.). *Water: A Comprehensive Treatise*. 7 Volumes. Plenum Press: New York, 1972–82.
2. J. G. Powles. *Adv. in Phys.* **22** (1973), 1.
3. L. Blum & A. H. Narten. *Adv. in Chem. Phys.* **34** (1976), 203. Narten has used a different notation in later publications.
4. P. A. Egelstaff. *An Introduction to the Liquid State*. Academic Press: New York, 1967, Chap. 5.
5. V. F. Sears, 'Thermal neutron, scattering lengths and cross sections for condensed matter research'. *Chalk River Report AECL-8490*, 1984.
6. L. Koester. *Springer Tracts in Modern Physics* **80** (1977), 1.
7. H. Bertagnolli & M. D. Zeidler. *Mol. Phys.* **35** (1978), 177.
8. J. Waser & V. Schomacher. *Rev. Mod. Phys.* **25** (1953), 671.
9. M. B. Deraman, J. C. Dore, J. G. Powles, J. H. Holloway & P. Chieux. *Mol. Phys.* To be published.
10. P. A. Egelstaff, J. A. Polo, J. H. Root, L. J. Hahn & S. H. Chen, *Chem. Phys. Rev. Lett.* **47** (1981), 1733.
11. L. Bosio, S. H. Chen & J. Teixeira. *Phys. Rev.* A **27** (1983), 1468.
12. J. E. Enderby & G. W. Neilson. In *Water*, Vol. 6 (ed. F. Franks). Plenum Press: New York, 1979, chap. 1.
13. L. van Hove. *Phys. Rev.* **95** (1954) 249.

14. G. Placzek. *Phys. Rev.* **86** (1951), 377.
15. C. G. Schull. 'Physics with early neutrons'. In '*Proc. Conf. on Neutron Scattering*', *Gatlinburg, USA* (*1976*): *Report* 760601 (ERDA), p. 1.
16. P. A. Egelstaff & M. J. Poole. *Experimental Neutron Thermalisation.* Pergamon Press: Oxford, 1970.
17. D. I. Page. In *Water*, Vol. 1 (ed. F. Franks). Plenum Press: New York, 1972, Chap. 9.
18. A. H. Narten & H. A. Levy. *J. Chem. Phys.* **55** (1971), 2263.
19. A. H. Narten. In *Water*, Vol. 1 (ed. F. Franks). Plenum Press: New York, 1972, Chap. 8.
20. H. H. Paalman & C. J. Pings. *J. Appl. Phys.* **33** (1962), 2635.
21. B. H. Meardon. *AERE Report* R7302, 1973.
22. The instruments currently installed are described in *Neutron Research Facilities at the ILL Reactor.* B. Maier (ed.), 1983.
23. P. Ageron. *Endeavour* 1972, p. 67.
24. J. P. Ambroise & R. Bellisent. In *Position Sensitive Detection of Thermal Neutrons.* (eds P. Convert and J. B. Forsyth). Academic Press: New York, 1983, p. 286.
25. Neutron sources based on electron linear accelerators were first used in England (AERE, Harwell) and Japan (Tohoku University).
26. C. G. Windsor. *Pulsed Neutron Scattering.* Taylor and Francis: London, 1981.
27. R. N. Sinclair, D. A. G. Johnson, J. C. Dore, J. H. Clarke & A. C. Wright. *Nucl. Instrum. Methods* **117** (1974), 445.
28. J. C. Dore & J. H. Clarke. *Nucl. Instrum. Methods* **136** (1976), 79.
29. G. Manning. *Contemporary Phys.* **19** (1978), 505.
30. Neutron sources based on proton accelerators have been developed in USA (Argonne Lab. and WNR, Los Alamos), Japan (KENS) and England (SNS, Rutherford–Appleton Lab.).
31. *Proc. Workshop on Neutron Scattering Instrumentation for SNQ* (eds R. Schern and H. Stiller), KFA Julich, Report, July 1954, 1984.
32. J. H. Clarke. LINDA-program for Linac neutron diffraction analysis. *AERE Report* R-8121, 1975.
33. J. R. Granada, G. W. Stanton, J. H. Clarke & J. C. Dore. *Mol. Phys.* **37** (1979), 1297 and earlier papers in the same series.
34. J. G. Powles. *Mol. Phys.* **36** (1978), 1181.
35. J. G. Powles. *Mol. Phys.* **37** (1979), 623 and earlier papers.
36. L. Blum, A. H. Narten & M. Rovere. *J. Chem. Phys.* **73** (1980), 3729.
37. S. A. Chen, J. Teixeira & R. Nicklow. *Phys. Rev.* A **16**, (1982), 3477 and references cited therein.
38. M. R. Chowdhury, J. C. Dore & J. B. Suck, to be published.
39. K. Skold, J. M. Rowe, G. Ostrowski & P. D. Randolph. *Phys. Rev.* A **6** (1972), 1107; J. R. D. Copley & S. W. Lovesey. *Rep. on Progr. in Phys.* (1974).
40. V. F. Turchin. *Slow Neutrons.* Israel Programme for Scientific Translations, 1965.
41. R. M. Moon, T. Riste & W. C. Koehler. *Phys. Rev.* **181** (1969), 920.
42. J. C. Dore, J. H. Clarke & J. T. Wenzel. *Nucl. Instrum. Methods* **138** (1976), 317.
43. J. Mayers, R. Cywinski, T. J. L. Jones & W. G. Williams. *RAL Report* 84–118 and earlier papers by W. G. Williams.

44. D. I. Page & J. G. Powles. *Mol. Phys.* **21** (1971), 901.
45. A. H. Narten. *J. Chem. Phys.* **56** (1972), 5681.
46. G. Walford. Ph.D. Thesis, Univ. of Kent (unpublished), 1975.
47. G. Walford, J. H. Clarke & J. C. Dore. *Mol. Phys.* **33** (1977), 25.
48. G. Walford & J. C. Dore. *Mol. Phys.* **34** (1977), 21.
49. I. P. Gibson, Ph.D. Thesis, Univ. of Kent (unpublished), 1978.
50. I. P. Gibson & J. C. Dore. *Mol. Phys.* **37** (1979), 1281.
51. N. Ohtomo, K. Tokiwano & K. Arakawa. *Bull. Chem. Soc. Japan* **54** (1981), 1802.
52. N. Ohtomo, K. Tokiwano & K. Arakawa.. *Bull. Chem. Soc. Japan* **55** (1982), 2788.
53. L. Bosio, J. Teixeira, J. C. Dore, D. C. Steytler & P. Chieux. *Mol. Phys.* **50** (1983), 733.
54. G. W. Neilson, D. I. Page & W. S. C. Howells. *J. Phys.* D **12** (1979), 901.
55. A. Y. Wu, E. Whalley & G. Dolling. *Chem. Phys. Lett.* **84** (1981), 433.
56. A. Y. Wu, E. Whalley & G. Dolling. *Mol. Phys.* **47** (1982), 603.
57. G. Neilson. AIRPT Meeting. *Mat. Res. Soc. Symp. Proc.* **22** (1984), 111.
58. J. H. Clarke, J. C. Dore & H. Egger. *Mol. Phys.* **39** (1979), 533.
59. D. I. Page & J. G. Powles. *Mol. Phys.* **29** (1975), 1287.
60. P. A. Egelstaff and A. K. Soper, *Mol. Phys.*; J. G. Powles. *Mol. Phys.* **42** (1981), 757.
61. W. S. C. Howells. *J. de Physique*: Suppl. C 7 (1984), 81.
62. J. H. Clarke & J. C. Dore, unpublished data.
63. A. K. Soper & R. N. Silver. *Phys. Rev. Lett.* **49** (1982), 471.
64. A. K. Soper. *Chem. Phys.* **88** (1984), 187.
65. J. C. Dore, W. S. C. Howells, R. N. Sinclair & A. C. Wright. SANDALS, 'A Small Angle Neutron Diffractomer for Amorphous and Liquid Samples. A design study for the Spallation Neutron Source.' Rutherford–Appleton Lab.
66. F. Hajdu, S. Lengyel & G. Palinkas,. *J. Phys.* D **12** (1979) 901.
67. R. W. Impey, M. L. Klein & I. R. MacDonald. *J. Chem. Phys.* **74** (1981), 647; O. Matsuoka, E. Clementi & M. Yoshimine. *J. Chem. Phys.* **64** (1976), 1351; G. C. Lie, E. Clementi & M. Yoshimine. *J. Chem. Phys.* **64** (1976), 2314.
68. A. V. Hobbs. *Ice Physics*. Clarendon: Oxford, 1974.
69. P. A. Egelstaff, D. I. Page & C. R. T. Heard. *J. Phys.* C **4** (1971), 1453.
70. S. A. Rice. *Current Comments in Chem. Phys.* 1976.
71. E. F. Burton & W. F. Oliver. *Proc. Roy. Soc.* A **153** (1935), 166; see ref. 68 for later work.
72. P. Brugeller & E. Mayer. *Nature* **288** (1980), 569 and **298** (1982), 715.
73. J. Dubochet & J. Lepault. *J. de Physique: Colloque C 7* (1984), p. 85.
74. O. Mishima, L. D. Calvert & E. Whalley. *Nature*, **310** (1984) 393.
75. M. G. Sceats & S. A. Rice. In *Water*, Vol. 7 (ed. F. Franks). Plenum Press: New York, 1982, chap. 2.
76. A. H. Narten, C. A. Venkatesh & S. A. Rice. *J. Chem. Phys.* **64** (1976), 1106.
77. J. T. Wenzel, C. V. Linderstrom-Lang & S. A. Rice. *Science* **187** (1975), 428.
78. M. R. Chowdhury, J. C. Dore & J. T. Wenzel. *J. Non. Cryst. Solids* **53** (1982), 247.
79. R. Boutron & R. Alben. *J. Chem. Phys.* **62** (1975), 4848.
80. D. E. Polk. *J. Non Cryst. Solids*, 5 (1971), 365.

81. F. H. Stillinger. *Adv. in Chem. Phys.* **31** (1975), 1.
82. M. R. Chowdhury, J. C. Dore & D. G. Montague. *J. Phys. Chem.* **87** (1983), 4037.
83. R. J. Speedy. *J. Phys. Chem.* **88** (1984), 3364.
84. L. Bosio, J. Teixeira & H. E. Stanley. *Phys. Rev. Lett.* **46** (1981), 597.
85. L. Bosio, J. Teixeira, J. C. Dore & P. Chieux, unpublished results.
86. S. Shibata & L. S. Bartell. *J. Chem. Phys.* **42** (1965), 1147.
87. E. Whalley. *Mol. Phys.* **28** (1974), 1105.
88. W. F. Kuhs & M. S. Lehman, J. Phys. Chem. **87** (1983), 4312.
89. W. F. Kuhs, private communication and *Proc. of Colston Symposium* (Bristol). 1985. To be published.
90. J. L. Finney. This volume, p. 93.
91. W. J. Jorgenson. *J. Amer. Chem. Soc.*
92. P. Barnes, J. L. Finney, J. D. Nicholas & J. E. Quinn. *Nature.* **282** (1979), 459.
93. R. A. Kuharski & P. J. Rossky. *Chem. Phys. Lett.* **103** (1984), 357.
94. B. Berne. In preparation.
95. P. A. Egelstaff & J. H. Root. *Chem. Phys.* **76** (1983), 405.
96. H. E. Stanley & J. J. Teixeira. *J. Chem. Phys.*, **73** (1980), 3404.
97. A. Geiger & H. E. Stanley. *Phys. Rev. Lett.* **49** (1982), 1749 and earlier papers.
98. H. E. Stanley, R. L. Blumberg & A. Geiger. *Phys. Rev.* B **28**, (1983), 1626.
99. A. Geiger, P. Mausbach, J. Schnitker, R. L. Blumberg & H. E. Stanley, *J. de Physique; Colloque C7* (1984), p. 13.
100. C. A. Angell & V. Rodgers *J. Chem. Phys.* **80** (1980), 6245.
101. F. Hajdu. *Acta Chim. (Budapest)* **93** (1977), 371.
102. G. Torchet, P. Schwartz, J. Farges, M. F. de Feraudy & B. Raoult. *J. Chem. Phys.* **79** (1983), 6196.
103. N. Ohtomo, K. Tokiwano & K. Arakawa. *Bull. Chem. Soc. Japan.* **57** (1984), 329.
104. A. Angell. In *Water.* Vol. 7 (ed. F. Franks). Plenum Press: New York. 1982, chap. 1.
105. E. Lang & H.-D. Lüdemann. *Ber. Bunsen Gesselschaft für Phys. Chem.* **84** (1980) 462.
106. Recent neutron diffraction measurements on a D_2O/emulsion system at -29 °C have been made by L. Bosio, M.-C. Bellisent, J. Teixeira and J. C. Dore.
107. G. P. Johar. *Phil. Mag.* **35** (1977), 1077; S. A. Rice, M. S. Bergren & L. Swingle. *Chem. Phys. Lett.* **59** (1978), 14.
108. J. G. Powles, E. K. Osae, J. C. Dore & P. Chieux. *Mol. Phys.* **43** (1981), 1051.
109. J. G. Powles, J. C. Dore, E. K. Osae, J. H. Clarke, P. Chieux & S. Cummings. *Mol. Phys.* **43** (1981).
110. M. Deraman, J. C. Dore, J. G. Powles, J. H. Holloway & P. Chieux. *Mol. Phys.* In preparation.
111. G. Palinkas, E. Kalman & P. Kovacs. *Mol. Phys.* **34** (1977), 525.
112. E. Kalman, G. Palinkas & P. Kovacs. *Mol. Phys.* **34** (1977), 505.
113. J. C. Dore & G. Palinkas. Comments at Faraday Disc. No. 66. *Structure and Motion in Molecular Liquids.* 1978, pp. 87–89.
114. J. G. Powles, J. C. Dore & D. I. Page. *Mol. Phys.* **24** (1972), 1025.
115. J. Reed, Ph.D Thesis, Univ. of Kent (unpublished), 1981.
116. W. E. Thiessen & A. H. Narten. *J. Chem. Phys.* **77** (1982), 2656.

117. A quantitative comparison of the separate data sets is crucial to further progress in this field but is by no means a trivial operation as it will require extensive computation and re-analysis of existing data in considerable detail; it is hoped that resources for this work can be found in the near future.
118. O. Bastiansen & M. Traetteberg. *Acta Cryst.* **13** (1960), 1108.
119. A. H. Narten, W. E. Thiessen & L. Blum. *Science* **217** (1982), 1033.
120. P. A. Egelstaff & J. H. Root. *Chem. Phys.* **76** (1983), 405.
121. J. C. Dore. *J. de Physique; Suppl. C 7* (1984), p. 57.
122. D. G. Montague, J. C. Dore & S. Cummings *Mol. Phys.* **53** (1984) 1049.
123. J. Janzen & L. S. Bartell. *J. Chem. Phys.* **50** (1968), 3611.
124. S. V. Cyvin. *Molecular Vibrations and Mean Square Amplitudes.* Elsevier: Amsterdam, 1968, chap. 14.
125. J. D. Bernal & R. H. Fowler. *J. Chem. Phys.* **1**, 515 (33).
126. D. C. Steytler, J. C. Dore & C. J. Wright. *Mol. Phys.* **48** (1983), 1031.
127. L. Blum & A. H. Narten. *J. Chem. Phys.* **64** (1976), 2804.
128. J. G. Powles. *Mol. Phys.* **36** (1978), 1181.
129. J. Teixeira, M.-C. Bellisent-Funel, S.-H. Chen & A. J. Dianoux. *Phys. Rev.* A **31** (1985) 00.
130. J. Teixeira & M.-C. Bellisent-Funel. Private communication. J. Teixeira. *Proc Colston Symposium, Bristol, 1985* (ed. G. Neilson). To be published.
131. S.-H. Chen & J. Teixeira. *Adv. Chem. Phys.* (1984). In press.
132. M.-C. Bellisent-Funel & J. C. Dore. Unpublished data.
133. J. C. Dore. *Proc. Colston Symposium, Bristol, 1985* (ed. G. Neilson). To be published.
134. W. Kuhs. *Proc. Colston Symposium, Bristol, 1985* (ed. G. Neilson). To be published.

A Further Postscript (added at proof stage)

A paper by Giguère has drawn attention to features in the shape of Raman spectrum for water (*J. Raman Spec.* **15**, (1984), 354) and suggested that many of the problems in interpreting the experimental data can be resolved by considering the presence of bifurcated hydrogen bonds. The bifurcated bond corresponds to an orientation in which the hydroxyl group is effectively hydrogen bonded to two neighbouring oxygen atoms as shown in figure A1. Within the overall tetrahedral network of oxygen atoms shown in figure 21 this gives an intermediate position in which the O—H distance is increased from 1.8 to 2.4 Å as shown in figure A2. Giguère concludes that the temperature-dependence of the Raman spectra can be explained by an increase of bifurcated bonds as the temperature is increased. This has immediate relevance to the structural characteristics revealed in the $\Delta d_L(r, T)$ function of figure 13 and provides an explanation (Giguère, private communication) of the shape in the 1.5 to 3.0 Å region due to changes in the relative populations. The negative peak at 1.75 Å represents a reduction in the number of linear hydrogen bonds and the positive peak at 2.8 Å represents an enhancement in the number of bifurcated hydrogen-bonds. This interest-

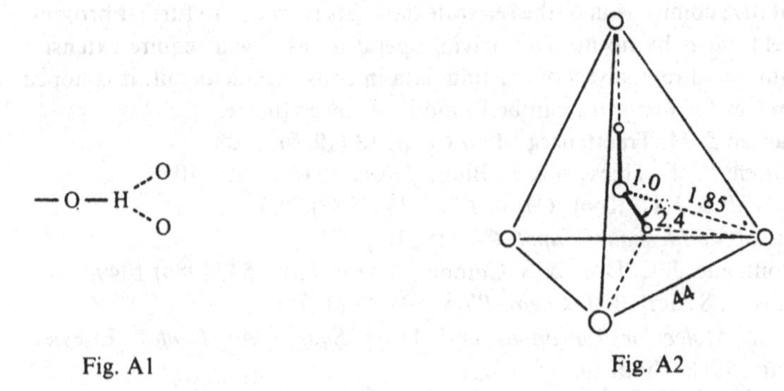

Figure A1. A bifurcated hydrogen bond.

Figure A2. Distance changes for a bifurcated hydrogen bond in a tetrahedral network [from Giguère].

ing speculation suggests that the breakdown of the hydrogen-bonded network with increasing temperature proceeds by a two-stage process of hydrogen-bond disordering from 'linear' to 'bifurcated' to 'broken' bonds. There is clearly a need for further study of the implications arising from this important conceptual approach to the geometrical characteristics of the network connectivity in water.

The Water Dimer Potential Surface

J. L. FINNEY, J. E. QUINN AND J. O. BAUM

Department of Crystallography, Birkbeck College, Malet Street, London WC1E 7HX, U.K.

In Memoriam: George Amer (1891–1985)

1. Introduction

The thermodynamics, dynamics and structure of any condensed phase depends ultimately on the interatomic or intermolecular interactions; an adequate theory of the liquid state would be able to calculate structure and properties at a given thermodynamic state point using only a knowledge of the elementary potential function. The structure of 'simple' liquids such as inert gases is determined essentially by the repulsive core of the potential function; the *details* of the structure can be considered as perturbations from an 'ideal' structure, which to zeroth order approximation relates to Bernal's random packing of hard spheres. [**1**–**4**]

The (relative) simplicity of 'simple' liquids is due mainly to the isotropic nature of the potential function Φ. Neglecting three-body effects – which for these systems can be considered as a perturbation – the potential energy of a pair of particles is determined solely by their separation *r*, so that

$$\Phi = \Phi\ (r \text{ only}). \tag{1}$$

This spherical symmetry, together with the sufficient hardness of its repulsive core, means that the structures of the condensed phases (solid and liquid) are determined largely by packing considerations. Thus, we can immediately explain qualitatively such 'normal' liquid behaviour as contraction on freezing, expansion on heating, and the increase in viscosity with pressure.

For molecular liquids, this simplicity is lost, as the potential function becomes a more complex function of relative position and orientation of a pair of molecules. The departure of the behaviour of molecular liquids from 'normality' will depend upon the nature of the intermolecular interactions. Most still behave 'normally' in terms of volume changes on freezing and expansion on heating. Water, however, is clearly different, although not unique, in that these properties do not follow the normally expected behaviour, and in fact several of these so-called 'anomalous' properties (e.g. expansion on freezing) are crucial for supporting life processes and maintaining their environment.

In so far as liquid properties depend ultimately on the elementary intermolecular interactions, it is reasonable to look for the origin of this 'anomalous' behaviour in the water–water potential function. In fact, we find it in the nature of the directionality of the water potential function, and conclude that such properties as volume expansion on freezing, and the existence of a temperature of maximum density should be observed in other liquids with similar molecular level interaction geometry. This is in fact the case, as is found with SiO_2.

However, unlike simple liquids, our understanding of the structure–potential function relationships in water is unclear. Although an idealized structure such as the random tetrahedral network model developed originally (in rather different ways) by Pople [**5**] and Bernal [**6**] is adequate to account qualitatively for the behaviour of the liquid, we have yet to understand how real water deviates from this ideal model. This problem is not one of irrelevant detail, as many of those aspects of water which are central to its role in chemistry and biology are not inherent in this simple model. Energy balances in biomolecular processes are extremely fine, and the currently unclear subtleties of the water interactions and their perturbations appear crucial in maintaining these balances. [**7, 8**]

With the advent of computer simulation techniques, much attention has been directed towards the derivation of potential functions for water, both for use in calculations on pure water itself, and in interaction with ions and molecules of chemical and biological interest. In what follows, we will discuss the derivation and testing of these potentials, with particular reference to their success or otherwise in explaining structural results drawn from experimental scattering measurements. Most of the discussion will be within the framework of rigid body models of water, avoiding problems of quantum modifications. This is largely because little work has been done outside these restrictions, rather than from a belief that such restrictions are reasonable: recent work suggests that an adequate modelling of water will require moving outside this restrictive framework.

The next section discusses briefly the water structure data available for comparison with predictions from assumed potential functions. The problems of obtaining these data experimentally are assessed in the accompanying article by Dore [**9**]; we restrict ourselves here to a discussion of the salient structural information contained in the pair correlation functions that will be used in later sections. Recent theoretical work on water-like molecules using analytical theories is then discussed to underline the gross features of the potential function that give rise to the 'anomalous' properties. The subsequent section describes a representative sample of dimer potentials for water; it is followed by an examination of the similarities and differences in their distance and orientational dependencies. We then assess the relative performance of various potentials in predicting liquid state structure, and conclude that none is adequate; possible underlying reasons for the failure of currently available potentials are discussed. Finally, we conclude with a discussion of possible ways forward, including previously neglected indications from both recent theoretical work on dimer potentials in general, and experimental work on related aqueous systems.

2. Pair Correlation Functions

In assessing the ability of proposed potential functions to explain the structure of the liquid, we will compare structural predictions with direct experimental structural data. We will therefore be interested in pair correlation functions (PCFs) which can be obtained both theoretically from computer simulation calculations using model potentials, and experimentally from scattering studies.

The pair correlation function $g(r)$ specifies the probability of finding an atom (or other specified centre) at a distance r from another atom at $r = 0$. A typical PCF for a simple liquid with an isotropic potential function is illustrated in figure 1 (broken curve). The interpretation of the function is fairly straightforward, with the peak *positions* indicating the most probable neighbour distances, the peak *widths* the dispersion of these distances, and the areas relating to the number of neighbours within a specified distance band. The first peak position corresponds to the nearest neighbour distance r_0, with the peak area (which is not clearly defined in the case shown, leaving several essentially arbitrary possibilities for defining the upper limit of integration) relating to the number of nearest neighbours. The second peak, which occurs at about 1.7–2.0 r_0 in the simple liquid case, can be explained geometrically in terms of possible structures determined by the packing constraints in a dense assembly of spherical atoms. [**2**, **10**]

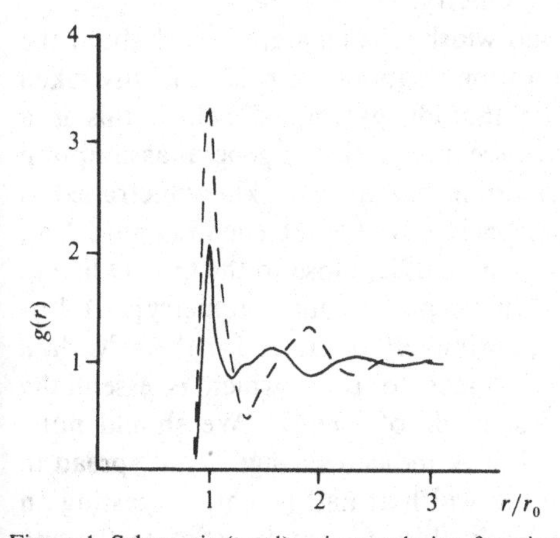

Figure 1. Schematic (total) pair correlation functions for liquid argon (broken line) and water (full line) [**12**] (scaled to give both first peaks at the same distance). Note that the status of the 'blip' in the water data at about $1.2r_0$ is unclear, and is generally accepted to be an artifact of the Fourier transform truncation.

For molecular liquids, the situation is complicated by the presence of two or more different types of atomic centre. A *total* PCF can still be defined, in which no distinction is made between the different atom centres in the molecule. Such highly averaged information is, however, of limited use. More useful are *partial* PCFs, in which the identities of the origin atom centre and its neighbours are specified. For the water case, we can define three partial PCFs, namely $g_{OO}(r)$, $g_{OH}(r)$ and $g_{HH}(r)$, where the subscripts identify the two atom types involved. Thus $g_{OH}(r)$ gives the probability of finding an H atom centre at a distance r from a central O. The experimental problems of extracting these partials are discussed by Dore in the accompanying article.[**9**] For our purposes, it is useful to discuss the nature of the configurational information in these partials before proceeding to examine the predictions of various potentials. We will assert for the present that $g_{OH}(r)$ and $g_{HH}(r)$ still remain to be experimentally established, while $g_{OO}(r)$ is known with greater confidence. However, we should bear in mind that there may still be a problem in the assumed known $g_{OO}(r)$ from X-ray diffraction, as the interpretation to date has generally assumed a spherical form factor for the O and H atoms in the H_2O molecule in liquid water.

Figure 1 also shows the form of the $g_{OO}(r)$ for water at room temperature and pressure (full line). For both PCFs, we observe the same general oscillatory behaviour, but with two major differences in the water case. First, the first peak has a smaller area than does the simple liquid, illustrating a lower coordination number of between 4 and 5, depending on the upper limit of the integration (cf. 8–10 for a simple liquid). This reflects the known lower number density of water compared to liquids whose structures are controlled largely by packing constraints. The other major difference is that the second peak has shifted to a much lower value of r, indicating a clear difference in the nature of the second-neighbour geometry.

This second peak position, area and width tells us a great deal about the average structure of the liquid. Its position at about 1.6 r_0 is generally taken as indicating a tetrahedral structure though, as argued below, this is a (misleading) oversimplification, which seems to reflect a general assumption of tetrahedrality in water–water interactions based on the known tetrahedral structures of the ices. The basic geometry involved is sketched in figure 2(a), which is drawn assuming the donor hydrogen lies close to the O···O line (a reasonable assumption from the known gas phase dimer geometry[**11**]). For a second peak centred at 1.63 r_0 (approximately the ratio for the 4 °C data of Narten *et al.*[**12**]) the 'ideal' θ_A comes to 106°, which is essentially indistinguishable from the tetrahedral angle of 109.47°. We should note, perhaps, that the width of the second peak indicates a significant spread in this angle, an estimate based on the full width at half height, suggesting an angular range of about 95–125°. Clearly this is a crude estimate (having ignored, for example, dispersion in r_0), but it does serve to underline the likely existence of a range of second-neighbour angles from about a right angle to

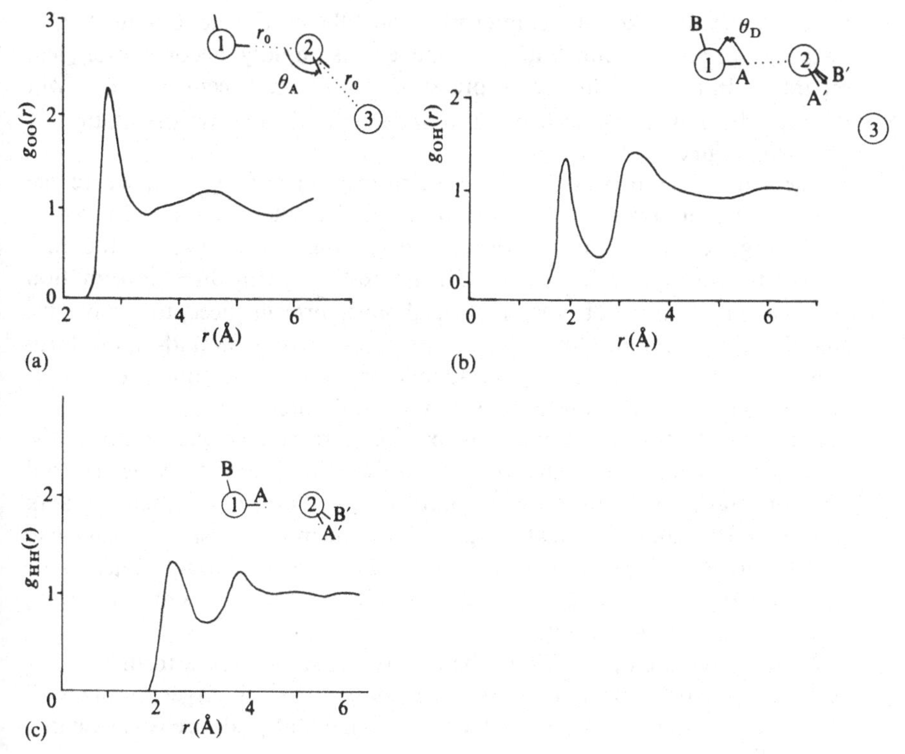

Figure 2. Schematic partial PCFs for a water-like system.

above the trigonal angle of 120°. $g_{OO}(r)$ thus gives good information on the orientational structure of second-neighbour oxygen centres of liquid water.

$g_{OH}(r)$ gives additional information on proton positions, and hence we might expect it to be highly sensitive to orientational structure. Consider, for example, the dimer configuration in figure 2(b) drawn to correspond to an ideal tetrahedral situation ($\theta_A = 109.47°$) with an O–O distance of 2.85 Å as in the liquid. Ignoring the *intra*molecular OH peak that will occur at about 1.0 Å ($r_{1A} \approx r_{1B} \approx r_{2A'} \approx r_{2B'}$), the first peak of *inter*molecular interest occurs at $r_{2A} \approx 1.85$ Å. The sharpness of this peak, after correcting for the spread in the r_{12} oxygen distance (known from $g_{OO}(r)$), indicates the degree to which the donor hydrogen lies off the O_{12} line. A second peak arises from r_{2B}, $r_{1A'}$ and $r_{1B'}$. Assuming ideal tetrahedrality, these distances will be almost equal (at about 3.25 Å). However, as the energy dependencies of θ_A and θ_D variations are different (see section 5), this second peak will represent a superposition of two distances of different structural relevance. A third feature at r_{3A} (ideally about 3.85 Å) will give some information on second neighbours, but in fact nothing additional to a good knowledge of g_{OO} and the first peak of g_{OH}. It will, however, be likely to overlap with the second peak, which will make a direct interpretation difficult. Further out, many OH distances begin to contribute and the effects rapidly become even less separable. In terms of direct interpretation, therefore, it seems that g_{OH} will probably be most useful in helping to understand the relative orientation of first neighbours.

The same general conclusion can be drawn concerning g_{HH}, where the number of overlapping HH distances rapidly becomes large as r increases. With a good knowledge of the first-neighbour peak of g_{OH}, the first intermolecular g_{HH} peak is particularly interesting, giving direct information on the orientation of the first-neighbour proton acceptor molecule (figure 2(c)). Beyond this peak, contributions from second- and third-neighbour molecules begin to overlap, and only a broad feature is seen in the simulated $g_{HH}(r)$ functions from pretty well all potentials.

An adequate model of water must be able to reproduce quantitatively the three partial correlation functions, when these have been fully determined experimentally. We emphasize the need for *quantitative* agreement of peak positions, areas and widths: although a successful model must have water-like PCFs as in figure 1, *prediction of this general* form is not enough. This is true of all three partials, though perhaps to different degrees, as the following discussion attempts to show.

Consider first g_{OO}, which probably gives the clearest information on second-neighbour correlations, being uncomplicated by larger numbers of OH or HH distances. Agreement with experimental peak areas is essential if coordination numbers (however defined) are to be correct: there is little point in having a correct form for g_{OO}, if there are too many or two few first (or higher) neighbours. The widths of the peaks give information on the dispersion of neighbour distances, which apart from being relevant to properties which depend upon fluctuations, may also be critical structurally (for example, too broad a first-neighbour distance distribution may 'wash out' the second-neighbour g_{OO} correlations). The precise positions of the peaks are also important. In particular, the position r_1 of the second g_{OO} peak is relatively insensitive to θ_A (figure 2(a)). Assuming for simplicity that $r_{12} = r_{23} = r_0$ in figure 2(a), we can write the r_{13}/r_0 ratio as $[2(1-\cos\theta_A)]^{\frac{1}{2}}$. From this we find

$$\frac{d(r_{13}/r_0)}{d\theta_A} = \frac{\sin\theta_A}{(r_{13}/r_0)};$$

as $\theta_A \sim 110°$, then the (r_{13}/r_0) ratio varies only slowly with changes in θ_A. We have already noted that the experimental width at half height of the second peak covers θ_A values from approximately 95° to 125°, a considerable spread when thought of in terms of local geometry. Similarly, r_{13}/r_0 changes of 5 and 10 per cent away from the ideal tetrahedral value correspond to average θ_A angles of 100° and 95° respectively. Clearly r_{13}/r_0 is an important structural parameter, for which *close* agreement with experiment is essential. Progression from average trigonal, through tetrahedral, to tetragonal coordination shifts r_{13}/r_0 over a range of only ± 10 per cent.

As explained above, the extractable second-neighbour information in g_{OH} and g_{HH} is limited by the overlap of a larger number of contributions beyond the first peak. This implies that comparisons between experiment and prediction must be made *very* critically indeed (and it also puts severe demands on the accuracy and precision of experimental determination): for example, changes in the profile of the second g_{OH} peak will reflect a different balance of contributions from $r_{1A'}$, $r_{1B'}$ and r_{2B} (figure 2(b)). Secondly, as will be seen in the next section, the *form* of g_{OH} is controlled largely by the first-neighbour overlap restrictions of the oxygen centres. The hydrogens are all attached to oxygens at an approximately constant distance: the first-neighbour peak of g_{OO} therefore implies g_{OH} correlations of the form of figure 2(b) even if there are weak – or even no – orientational restrictions on hydrogen positions (see, for example, figure 6(b)). When making comparisons through g_{OH}, therefore, we would expect all models with a reasonable g_{OO} first peak to give a g_{OH} of the form of figure 2(b); hence critical comparisons must be made of the peak shapes and areas. Many authors have noted previously that model g_{OH} and g_{HH} correlation functions are all 'water-like' and very similar in *form*; we would underline here, on the basis of the above discussion, that the *form* for all model potentials with the hydrogen attached to an oxygen will *of necessity* be 'water-like'. Comparisons using these partials must therefore be made at a quantitatively even more critical level than those made with the aid of g_{OO}: important aspects of the structure will be evident in the quantitative *details* of g_{OH} and g_{HH}, and comparison must be made at this level.

Partly for these reasons, and also because of uncertainties associated with the experimental determination of g_{OH} and g_{HH},[**9**] we restrict most of our comparative discussions to g_{OO}.

3. Theoretical Work: Essential Features of the Potential Function

There is an extensive theoretical framework applicable to *simple* liquids, which aims to derive the pair correlation function and thermodynamic properties (e.g. internal energy, compressibility) from an assumed knowledge of the potential function.[**13–15**] Such methods, despite the approximations necessary because of mathematical intractability, give useful results in many situations. However, they are not easily extended to molecular liquids like water, whose potential functions are strongly orientation-dependent. Thus, nearly all work on water uses computer simulation techniques [**16**] to derive structural and dynamic properties are consequent upon particular assumed potential functions. The application of such techniques to water has been reviewed by Wood.[**17**]

Before beginning to discuss water potential functions in detail and their successes and defects as revealed through simulation calculations, we mention here recent work using theoretical approaches. Despite the drawbacks of necessary approximations, recent analytical theories do give an interesting insight into the possible consequences for structure and properties by making systematic changes in certain aspects of a potential function. Similar information can, of course, be obtained from computer simulation with less uncertainty, but at *much* greater computational cost. Hence, despite the uncertainties consequent upon the approximations in the theories, the results of such work can be illuminating, and can suggest useful directions for future development of potential functions.

3.1. *Bernal and Fowler*

The first attempt to estimate the properties of water from an assumed interaction was made by Bernal and Fowler in 1933.[**6**] Their water model interestingly avoided the assumption of strong tetrahedrality, but instead related closely to more recent potentials based on a trigonal distribution of charge. The absence of any liquid state theory at all forced them to use qualitative arguments, based on analogy with known silicate structures; despite its qualitative nature, this work argued what we now know to be the essential nature of the water–water interaction, and the essential structure of the liquid.

3.2. *Percus-Yevick: Ben-Naim*

Some indication of the structural consequences of strong potential function directionality was given over 10 years ago by Ben-Naim, using a perturbation approach and an approximate version of the Percus–Yevick equation.[**18**] The pair-potential of two trial 'water-like' particles with relative positions described in terms of their intermolecular distance r_{12} and a set of angular variables $\mathbf{\Omega}$, was written as

$$U(r_{12}, \mathbf{\Omega}, \lambda) = U_{\mathrm{LJ}}(r_{12}) + \lambda S(r_{12})\, U_{\mathrm{HB}}(r_{12}, \mathbf{\Omega}). \tag{2}$$

In this equation the potential is split into a spherically symmetric term (here assumed of Lennard–Jones form), plus an angular-dependent perturbation term whose contribution can be increased through the variable coupling parameter λ.

Using a form for the 'hydrogen bonding' part U_{HB} derived from the simple Bjerrum model, based on equal positive and negative charges at the corners of a regular tetrahedron (the Ben-Naim–Stillinger, or BNS potential, see section 4.1.1 (*c*) and figure 9), the effect of increasing the coupling parameter, and hence the contribution of the angular dependence of the potential, can be studied. Convergent solutions of the integral equation used could be found only for relatively small λ values ($\lambda \leqslant 0.3$), but even for this reduced degree of orientational dependence the results are interesting. Figure 3 shows the

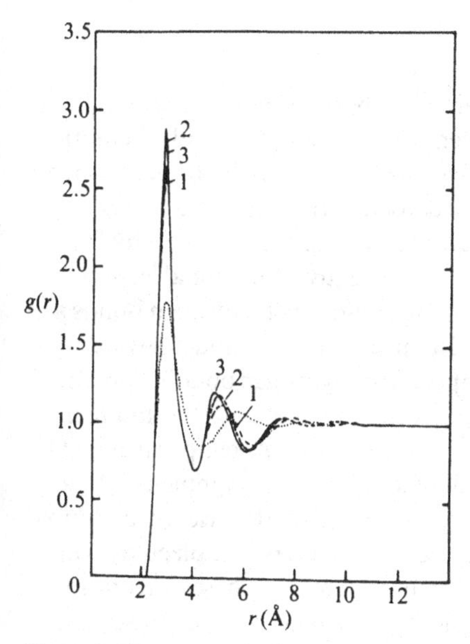

Figure 3. The response to increasing λ in eqn (2) (and hence increasing tetrahedral orientational dependence of the potential function). The dotted line is the reference $\lambda = 0$ curve.

results for the pair correlation function $g(r)$. For $\lambda = 0$, we have the normal simple-liquid-like function (though with a reduced first peak reflecting the lower density), with the second peak occurring at approximately $2r_0$. As the coupling parameter increases, two major changes occur. First, the first peak sharpens significantly, and moves to slightly lower r (presumably a consequence of the additional attraction). Secondly, and particularly interestingly, the second peak moves to much lower r. For $\lambda = 0.3$ the peak is at 4.8 Å, quite close to (but still greater than) the 4.6 Å position observed experimentally for water (although r_{13}/r_0 at 1.72 is too high, implying an O···O···O angle of about 120°).

Thus, we can conclude that adding a tetrahedral orientational dependence to an underlying spherically symmetric potential function results in changes in the gross features of the pair correlation function that begin to make it look water-like in terms of the position of the second peak. We should underline, however, that we are considering here only the gross features. Ben-Naim stresses that other features of the computed PCFs are in disagreement with experimental results for water, and instances the average co-ordination of about 7.4, compared with an experimental value of about 4.4. To this can be added the greater heights, compared with experiment, of both the main peaks; although we should bear in mind the approximate nature of the calculation, this excessive height in the first peak is a common feature of most simulated water g_{OO}s, to which we return in section 6.

3.3. *Cluster Theories: Andersen*

Since this early work of Ben-Naim, there have been more recent attempts to study water-like systems through extensions of simple-liquid theoretical techniques. Andersen [**19**, **20**] developed a cluster theory of hydrogen-bonded fluids, which has recently been extended to describe fluid water. [**12**] The theory is cast in terms of a reference potential and a sum of 'hydrogen bonding' interactions between positive and negative functional groups on each molecule; approximations to the average number of hydrogen bonds are then derived from the cluster theory. In specific application to water-like particles, Dahl and Andersen [**21**] decomposed the hydrogen bonding potential into a 'strong' and 'weak' part, and proceeded to calculate the number of strong and weak 'hydrogen bonds' as a function of temperature. The simplified potential they used allowed unambiguous assignment of non-bonded, strongly and weakly bonded pairs through the use of a double square-well potential, which describes the energy of two molecules whose relative orientations are within given limits (figure 4). These orientational limits, together with the distance and energy parameters, were chosen to be consistent with the qualitative behaviour of the same strongly tetrahedral BNS potential (see section 4.4.1 (*c*)) used previously by Ben-Naim.

The particular interest of this work is the simplicity of the potential, and its relative success in explaining *qualitatively* several 'anomalous' properties of water. For example, a density maximum and compressibility minimum were obtained, together with a large heat capacity, which was explained in terms of the possibility of distortion of the hydrogen bonds away from their optimal geometry without complete breakage (strong→weak bonds in the authors' terminology). The density maximum results from competition

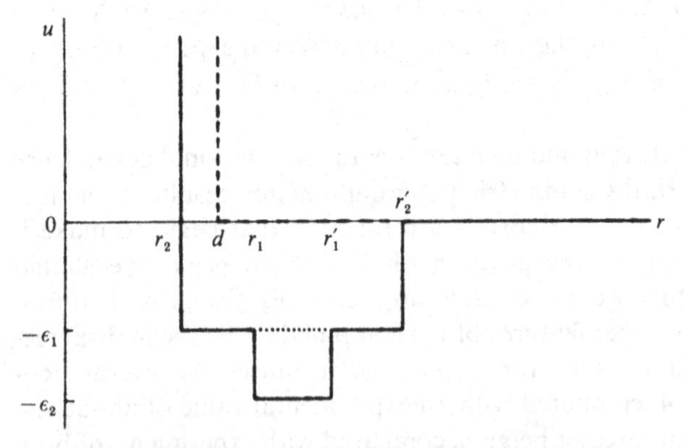

Figure 4. Orientation-dependent model potential used by Dahl and Andersen. The solid curve is for orientations leading to 'strong hydrogen bonding', and the dotted curve for orientations leading to 'weak hydrogen bonding'. The broken curve is for other orientations which lead to no hydrogen bonding. Reproduced, with permission, from reference 21.

between directionally dependent hydrogen-bonding forces driving towards an open (here tetrahedral) arrangement, and the short-range repulsive forces favouring a more close-packed structure.

These results strengthen the widely accepted association between the so-called 'anomalous' properties of water and the orientational dependence of the potential function. As the authors stress, however, the crudity of the model cannot be expected to give quantitative agreement. For example, their calculations gave no indications of known anomalous behaviour in the supercooled region. [**22**] Secondly, they were not able to change the potential function parameters in such a way as to improve *together* the density maximum, energy and density: parameter changes could improve any one of these, but only at the expense of worsening the agreement of the other two properties. We might tentatively conclude, therefore, that although qualitatively explaining several of the interesting properties of water (in a way that generally agrees with our previous concepts), the water interaction model used is probably too crude to be able to explain quantitatively the behaviour of liquid water.

3.4. *Integral Equations: Pettitt and Rossky*

A second theoretical attack recently developed by Pettitt and Rossky [**23**] uses an integral equation approach based on an extension of the RISM theory of Chandler and Andersen. [**24**] Several three-site models of the water molecule are used in which Lennard–Jones and charge sites are placed at the oxygen and hydrogen nuclei. Thus, no explicit tetrahedrality is assumed in the orientational dependence of the potential function: all orientational behaviour results directly from the sum of the various site–site interactions.

We consider here the results of this work with reference to the SPC (simple point charge) water model of Berendsen *et al.* [**25**] (section 4.4.1), and variants thereof. Comparing the g_{OO} PCF from computer simulation with that predicted from the theory shows a $1.6r_0$ peak in both cases; in the theoretical PCF, however, its intensity is too low (figure 5). Moreover, there is still a remnant of the $2r_0$ simple liquid peak, which is clearly absent in the simulation. It seems that the theory may not be able to handly *fully* the structural consequences of the orientational parts of the potential function. A further inadequacy in the theory is seen in the too high and sharp 'hydrogen bonding' g_{OH} first peak, which is also at too short a distance. Nevertheless, the correct g_{OO} qualitative features are seen, making it worthwhile to examine changes in the PCFs as modifications are made to the potential.

The basic model used in this work has a Lennard–Jones 6:12 potential centred on the oxygen, with a small repulsive r^{-12} coefficient related to the O···H intermolecular separation for technical reasons. No specific H···H repulsion or dispersion attraction is allowed for. The HOH angle is taken to be 109.48° rather than the experimental 104.5°, with an R_{OH} distance of

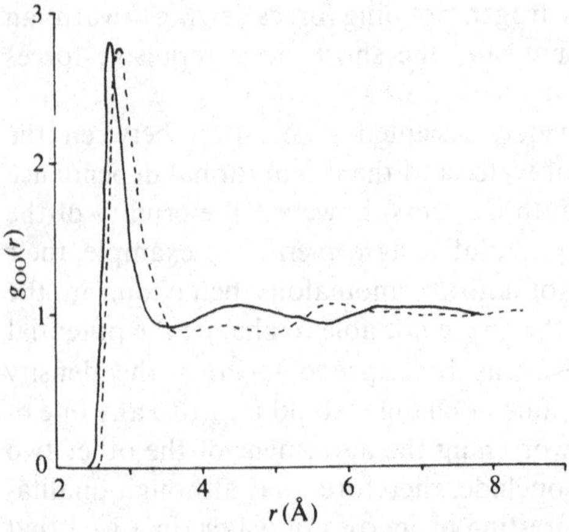

Figure 5. $g_{OO}(r)$ of essentially the SPC three-point charge model from the extended RISM theory[23] (broken curve), compared with the full simulation results[25] (full curve). Reproduced, with permission, from reference 23.

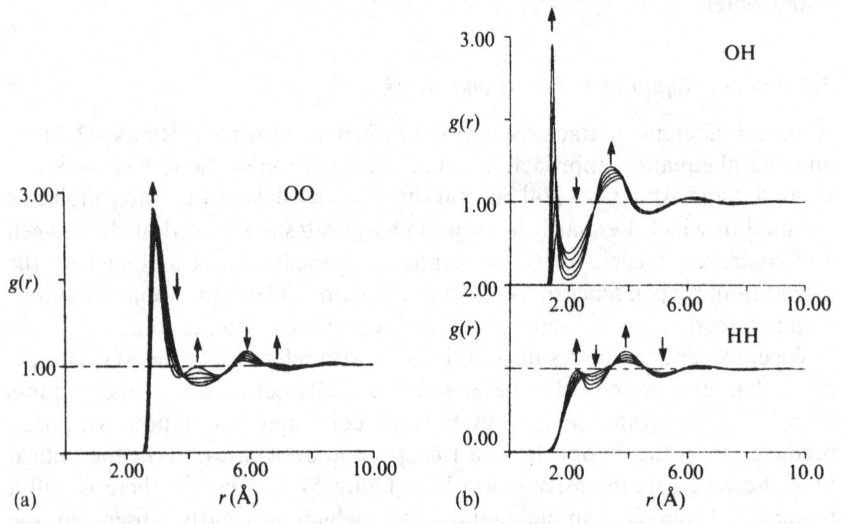

Figure 6. Partial correlation functions for essentially the SPC model[25] with varying site charge, according to the extended RISM theory of Pettit and Rossky.[23] Values of Z_H are 0.25, 0.30, 0.35, 0.40, 0.45e. Arrows indicate direction of change for increasing charge. N.B. Z_H values given here are correct; there is an error in the original paper (P. J. Rossky, personal communication). Reproduced, with permission, from reference 23.

1.000 Å. In the SPC model itself, the hydrogen charge is taken as 0.41e, with a compensating $-0.82e$ on the oxygen.

Figure 6 shows the changes that occur to the three partial correlation functions as the hydrogen charge Z_H is increased to 0.45e. We consider first the $Z_H = 0.25e$ functions; for this case, with only a weak r^{-12} repulsion between oxygen and hydrogen centres, the g_{OO} is essentially that of a simple liquid (see figure 6(a)). The g_{OH} will also be related to the simple liquid case, but including also the superposition of a free rotation, and thus containing little in the way of orientational structural information. Even so, we should note that g_{OH} (figure 6(*b*)) is water-like in having a first peak at about 1.6 Å and a second at about 3.7 Å. Thus even though this assembly has no orientational correlation specifically built into it, its qualitative appearance is what we have come to consider as water-like. This illustration underlines the point made in section 2 that *qualitative* similarity is totally inadequate when making comparisons between two pair correlation functions; we must make such comparisons very critically, considering the details of peak position, height and area, in so far as the latter can be defined for incompletely separated peaks. The g_{HH} form (figure 6(c)) is also similar to a water-like g_{HH} although the failure to resolve the first peak shows the weakness of the HH correlation.

As we 'charge up' the molecule, the g_{OO} becomes more water-like, as occurred with the earlier calculations of Ben-Naim, mentioned previously. As figure 6(a) shows, the $1.6r_0$ features grows – at the expense of *both* the $2r_0$ and r_0 peaks – while the first peak sharpens and moves to a slightly lower r. Bearing in mind the inability of the theory to reflect fully the consequences of increased orientational dependence of the potential function, we conclude that the potential function directionality consequent upon a reasonable three-point charge model is essentially capable of reproducing water-like behaviour in g_{OO}. We should underline that this potential function orientational dependence is not tetrahedral, but rather trigonal. Hence, water-like behaviour of g_{OO} is *not* dependent on a tetrahedral potential function, and for this reason alone the $1.6r_0$ peak should *not* be labelled – as is often done – as a 'tetrahedral' peak. Rather, it should be considered as a consequence of strong directionality of a geometry which leads to open structures – examples of which include both trigonality and tetrahedrality. We might recall in this context the small difference ($\sim$ 10°) between trigonal and tetrahedral angles, and also that the full width of the second peak in g_{OO} spans O···O···O angles from below 90° to above 120°.

A further point made by Pettitt and Rossky is illustrated in figure 6(a) and also seen in figure 1. This relates to the observation that beyond the first peak the g_{OO}s of water-like and simple liquids are out of phase: the second peak in the water case falls at a minimum of the scaled simple liquid g_{OO}, and vice versa. Thus, in trying to change progressively a potential from a simple liquid to a water-like type, the consequences for the g_{OO} PCF will be to reduce initially both the height of the $2r_0$ peak and the depth of the first minimum. A likely consequence at some point during the transition is a featureless g_{OO} beyond the first peak. However, this of itself does not necessarily indicate a structureless fluid – indeed, the presence of a strong first peak means that there must be structure at higher r; lack of PCF oscillations may result from the cancellation of simple liquid and water-like structural features in a liquid 'in transition'. The structure is there in the liquid, but is lost in the averaging in g_{OO}. A similar situation is well known in molten salts, where the g_{++} and g_{+-} tend to be out of phase; it is only when the partials are separated that the structure of the molten salt is revealed.[26] If one could label certain molecules as water-like and simple-liquid-like from their environments, similarly constructed partial PCFs would presumably show both water-like and simple liquid-like behaviour.

It seems useful to bear this point in mind throughout this discussion. As we shall see, it is unlikely that any of the hypothetical water potentials we shall discuss will fail to give some water-like behaviour, yet they may show featureless – or weakly oscillatory – g_{OO}s beyond the first peak (e.g. TIPS, see section 6) or the remnants of simple liquid features. It may be that this variety of behaviour is reflecting different degrees of orientational dependence of the potential function, and getting the answer 'right' may require a difficult, subtle balancing of these two characteristics.

As we increase Z_H, g_{OH} shows a shift to stronger orientational behaviour, the PCF (figure 6(b)) becoming what is thought to be more water-like. The first peak increases and narrows (probably too much, but this is an expected defect of the theory) while the second peak, which contains information on first-neighbour relative orientation, increases and moves to lower r. The first peak also shifts in, presumably as a direct consequence of the increased $O\cdots H$ attraction. As discussed in section 2, the second broad peak is likely to be the superposition of three OH distances involving both first- and second-neighbour waters (figure 2(b)), and hence it is dangerous to try to estimate the average tilt angle θ_A from this g_{OH}. Such a calculation would also be complicated by an apparent inconsistency in the position of the first g_{OH} peak with that of the first g_{OO}, presumably an inadequacy in the theory: for a g_{OO} first peak at 2.90 Å in figure 6(a) and an r_{OH} intramolecular distance of 1.0 Å, g_{OH} cannot be at a distance less than 1.90 Å; it is 1.41 Å in figure 6(b) for $Z_H = 0.45e$. We can perhaps say little more than that the position of the second g_{OH} peak is consistent with the generally accepted minimum-energy dimer structure, with the acceptor angle θ_A varying around 100–130° or so. It is difficult to interpret the data further than this.

Finally, upon charging, the g_{HH} distribution gains structure (figure 6(c)), becoming more apparently water-like. Pettitt and Rossky stress this is largely a consequence of the HH repulsion arising through the intramolecular correlations combined with the OH attraction.The precise behaviour of g_{HH} is thus likely to be *very* sensitive to the choice of parameters.

3.5. *Some General Comments*

Although none of these three theoretical approaches can be considered quantitatively sufficient, they do raise several points relevant to any discussion of potential functions with orientational dependence. First, even crude potential functions give a basic understanding of so-called anomalous properties of water, though one should perhaps not take too literally a potential function which (as in Dahl and Andersen's work [21]) has only three levels, and hence can be easily discussed in terms of strong, weak and non-existent 'hydrogen bonds'. We could perhaps note that such an approach was both insufficiently flexible to obtain a reasonable balance between the 'intensities' of the main properties considered (density maximum, energy and density), and gave no clue as to the nature of supercooling anomalies. Secondly, the $1.6r_0$ peak that is characteristic of water-like behaviour can be reproduced qualitatively by potential function directional dependencies of different types: specific tetrahedrality is not required, and so it is misleading to label the $1.6r_0$ peak as tetrahedral. We suspect that any orientational dependence in the potential function that leads to open structures (i.e. discriminates against O···O···O angles of less than 90°) will lead to similar behaviour. The orientational dependencies of three-point charge models are adequate to do this, the precise form of the resulting correlation functions probably depending subtly on the balance between competing effects. Thirdly, the fact that, beyond the first peak in g_{OO}, water-like and simple liquid oscillations are out of phase raises problems in making direct structural assertions from the relative strengths of the two features ($1.6r_0$ and $2r_0$ peaks) in g_{OO}s computed from different potential functions. Even complete flatness in this region cannot mean structureless if the first peak is strong, but rather a coincidental cancellation of the features arising from the two structural tendencies. This also underlines the need for comparisons between computed and experimental g_{OO}s to be made critically. It is not enough for a model to give the right *qualitative* features: the *quantitative* comparisons must be good if we are to accept our model as describing adequately the important, but subtle, structural aspects of the liquid which result from the interplay of orientational and distance dependencies in the potential function. In this context, temperature dependence of the PCFs becomes particularly important, as increasing temperature will change the balance away from water-like to simple liquid-like behaviour. As many biological processes are very sensitive to temperature, it would seem crucially important in such studies for us to reproduce temperature-induced structural shifts. The data are available [9], but little has been done yet to test potential functions in this way.

Finally, we should note the problems of using both g_{OH} and g_{HH} in understanding the geometrical nature of the underlying structure. We saw above that for the least charged model ($Z_H = 0.025e$), both g_{OH} and g_{HH} might be regarded as water-like, yet they contained very little in the way of orientational information. It is the *precise* positioning, and widths, of the various peaks which contain the orientational information, and this is in general interpretable only in conjunction with g_{OO}. We have already seen (section 2) that the basic features of both g_{OH} and g_{HH} arise from the O–O repulsion controlling the short-range oxygen order, which itself will define a minimum intermolecular OH distance. It is then the *differences* between the basic 'orientationally averaged' g_{OH} or g_{HH} and those obtained using a trial potential function which give us the useful information about the orientational structure of a particular model fluid under consideration. Thus, we stress again the need to be highly critical of our comparisons in the cases of all three partials; there are also very severe demands put on the accuracy and precision which we need from experimental measurements.

3.6. *Dipolar/Quadrupolar Fluids: Patey and Valleau*

Further interesting relationships between fluid structure and orientational dependence of potential functions come from work on dipolar–quadrupolar systems by Patey and Valleau.[27] They performed an extensive set of both thermodynamic perturbation theory and Monte Carlo calculations on a fluid of hard spheres with embedded quadrupoles, and additionally fixed dipoles. For our purposes, the major results from this work relate to the role of the quadrupole in determining the structure of the fluid.

Referring to their computed pair correlation function, a distinct tendency to suppress the simple-liquid $2r_0$ peak in favour of an increase in the 1.4–$1.8r_0$ region was noted. This tendency was underlined by the average electrostatic pair interaction between molecules about $2r_0$ apart being positive, clearly indicating an unfavourable (average) angular correlation at this distance, whether or not a dipole was present. Thus, it appears that the very existence of a significant quadrupole will tend to drive the structure away from one with a simple liquid-like PCF.

Patey and Valleau also concluded that quadrupoles have much greater structural consequences than 'comparable' dipoles (the comparison being made in terms of reduced units). For the quadrupolar system, they further interpret this in terms of a much broader range of angles which lead to strongly favourable interactions, and hence argue that quadrupolar molecules are able to 'cooperate' much more easily with neighbours in forming networks of favourable short-range interactions. Similar 'structural cooperation' is not possible for dipoles alone, where a significant number of intermolecular interactions will, of topological necessity in a condensed system, be repulsive. Quadrupoles are thus much more effective in orientationally ordering a dense system.

These comments are clearly relevant to water and aqueous solutions; they shift attention from the dipole to the correctness of the quadrupolar part of the potential function, and at the very least suggest that a hard sphere plus dipole is a totally inadequate model for water, unable even to approximate the angular correlations in the fluid. Table 1 shows the relative contributions to the total electrostatic configuration energy of Patey and Valleau's hard-sphere assemblies from the various dipolar and quadrupolar terms, for ratios of (reduced) moments relevant to water. For the value corresponding to the isolated water molecule dipole moment, we see that about two-thirds of the energy comes from the quadrupole–quadrupole term, the dipole–dipole

Table 1. *Relative contributions to the total electrostatic energy of simulated hard-sphere liquid assemblies for various quadrupole/dipole ratios (from reference 27)*

Q^{*2}/μ^{*2}†	1.32	1.11	0.88	0.71
Equivalent μ‡ for water quadrupole (D)	1.85	2.0	2.25	2.50
$E(\mu\mu)/E$ §	0.06	0.08	0.12	0.17
$E(\mu Q)/E$ §	0.32	0.37	0.39	0.41
$E(QQ)/E$ §	0.62	0.56	0.48	0.41

† Reduced values defined as $\mu^{*2} = \mu^2/kTR^3$, $Q^{*2} = Q^2/kTR^5$, where R is the hard-sphere diameter.

‡ Equivalent dipole moment for a quadrupole moment of approximately the water molecule value. 'Hard-sphere' diameter taken as 2.85 Å.

§ Fraction of configurational electrostatic energy contributed by the dipole–dipole ($\mu\mu$), dipole–quadrupole (μQ), and quadrupole–quadrupole (QQ) terms.

contributing only a small fraction (6 per cent). As the dipole moment is increased, the QQ term falls at the expense of the $\mu\mu$ and μQ terms, but even for a 35 per cent enhanced dipole moment (approximately the average enhanced value from polarizable model calculations of Barnes *et al.* [**28**, **29**]) the dipole–dipole term remains small. Although these figures are not quantitatively correct for water, they are relevant to water on two counts. First, they underline the importance of the quadrupole in determining the structure of a water-like assembly. Secondly, the balance between the various contributions changes significantly with increasing dipole moment. This is particularly relevant to water, where we know the dipole moment in condensed phases is enhanced by cooperative effects by, on average, 30–40 per cent, with a significant dispersion in individual molecule dipole moment values. Thus, as cooperative effects are allowed to be expressed, the relative energy contributions from the dipole and quadrupole will change, with possibly significant consequences for orientational structure. Moreover, at the local level, variations in dipole moment will result in different local contributions from the different energy terms; this local variation in dipole/quadrupole balance is likely to produce variations in local structure. This suggests that ignoring the dispersion in dipole moment resulting from cooperativity may affect local structure in a way which an effective pair potential (see section 4.1) may be inherently unable to handle.

3.7. *Summarizing Remarks*

The theoretical work discussed above serves to underline the importance of the orientational part of the potential function to both the structure of liquid water-like systems, and the anomalous properties of water. Although quite crude directional dependencies can reproduce some aspects of water-like behaviour, to explain structure and properties quantitatively requires much more subtle potential function prescriptions. The 'charging-up' of models of essentially trigonal symmetry underlines the fact that tetrahedrality is not essential for obtaining water-like behaviour. The simple dipole–quadrupole systems investigated show how important the quadrupole is to the building of energetically favourable three-dimensional networks, and point to a possibly significant change in the balance of the various contributions to the total electrostatic energy as the dipole moment is enhanced by cooperative effects. With this background, we now proceed to examine the design of specific potential functions, and their performance in simulations of liquid water.

Water Potential Functions

4.1. *Formal Statement*

We formally define an N-particle configuration of an assembly of molecules by a vector:

$$\mathbf{X}^N \equiv [\mathbf{X}_1, \mathbf{X}_2, \ldots \mathbf{X}_N] \tag{3}$$

where $\mathbf{X}_i$ depends upon both position $\mathbf{R}_i$ and orientation $\mathbf{\Omega}_i$ vectors. The configurational energy $U(\mathbf{X}^N)$ of the assembly can be written as an expansion in terms of interaction of pairs, triplets, etc., of molecules:

$$U(\mathbf{X}^N) = \sum_{i<j}^{N} U_2(\mathbf{X}_i, \mathbf{X}_j) + \sum_{i<j<k} U_3(\mathbf{X}_i, \mathbf{X}_j, \mathbf{X}_k) + \ldots, \tag{4}$$

where U_2 is the dimerization energy relative to the energies of the isolated molecules:

$$U_2(\mathbf{X}_i\,\mathbf{X}_j) = U(\mathbf{X}_i, \mathbf{X}_j) - U(\mathbf{X}_i) - U(\mathbf{X}_j). \tag{5}$$

U_3 specifies a three-body energy

$$\begin{aligned} U_3(\mathbf{X}_i, \mathbf{X}_j, \mathbf{X}_k) = {} & U(\mathbf{X}_i, \mathbf{X}_j, \mathbf{X}_k) - [U(\mathbf{X}_i) + U(\mathbf{X}_j) + U(\mathbf{X}_k)] \\ & - [U_2(\mathbf{X}_i, \mathbf{X}_j) + U_2(\mathbf{X}_i, \mathbf{X}_k) + U_2(\mathbf{X}_j, \mathbf{X}_k)]. \end{aligned} \tag{6}$$

Higher terms relating to four, five, etc. body energies can be defined analogously.

The above development is perfectly general; however, two simplifications are often made. First, in most models of water, the configurational energy of a molecule is assumed constant, i.e. $U(\mathbf{X}_i) = U(\mathbf{X}_j)$ for all i, j. There are at least two possible problems with this assumption. First, as intermolecular scattering may not in general be elastic, there is the possibility of exchange of energy between inter- and intramolecular motions. Secondly, the implicit assumption of the water molecule as a rigid body totally ignores intramolecular quantum effects, including the role of zero point motions of internal degrees of freedom. We may note also in passing that the classical approach almost universally taken ignores zero point vibrations with respect to intermolecular modes, which again may not be insignificant. Throughout the rest of this discussion, quantum effects are largely ignored, though recent work suggests greater attention should be given to these aspects of the problem.

The second assumption usually made is to neglect three-body and higher-energy contributions. This assumption is commonly made in simple-liquid studies, where the effects are generally small, and can be allowed for as a perturbation. For water, however, there is general agreement that three-body effects at least are *not* insignificant, several quantum mechanical calculations suggesting they can contribute up to 20 per cent of the energy of an assembly.[**30–32**] As this is of the order of $2kT$ at room temperature, it seems dangerous at first sight to ignore effects of this magnitude when estimating ensemble average properties.

Two ways have been proposed to handle this problem. The first tries to account for three-body (and higher) effects through the presence of a dipole polarizability term, as in the potentials developed by Campbell[**33**], the Birkbeck group[**28**, **29**, **32**, **34**], Berendsen[**35**] and Stillinger and David.[**36**] Such methods automatically account for four and higher body corrections, though these appear to be small.[**32**] Computational demands have prevented wider examination of polarizable potentials for water, though their use in molten salt calculations has been more extensive. Recent work shows that the slowly varying nature of the dipole moments in simulated assemblies allows approximations to be made which reduce the computational overheads to an almost negligible level.[**37**] Non-pair-additive potentials based upon the coordinates of triplets of molecules have been discussed, though the problems of both designing and using such a twelve-dimensional function have prevented any significant progress so far along these lines.

The second approach is to argue that an 'effective pair potential' (EPP) can be designed such that the microscopic structural and thermodynamic effects of the higher-order terms are taken into account. Formally, the nature of such a fitting has been developed by Stillinger,[**38–40**] who shows that a variational criterion allows the optimal assignment of a sum of effective pair potentials to any given non-pair additive function. This assignment causes the effective pair potential to differ from the actual two-body contribution in equation (4) so that it creates essentially the same structural shifts at equilibrium that would be produced by the additional three-, four- and higher-order terms in equation (4). The specific choice of the effective potential $U_{2\text{eff}}$ formally involves a minimization of

$$\int \cdots \int \left\{ \exp[-\tfrac{1}{2}U(\mathbf{X}^N)/kT] - \exp\left[-\frac{1}{2} \sum_{i<j}^{N} U_{2\text{eff}}(\mathbf{X}_i, \mathbf{X}_j)/kT \right] \right\}^2 d\mathbf{X}^N \tag{7}$$

from which we cannot expect to be able to extract an exact solution. An EPP chosen subject to this condition will simultaneously eliminate first-order errors in the Helmholtz free energy, singlet, and pair distribution functions; this statement does not apply to triplet and higher-order distribution functions.

Before proceeding to discuss the derivation of potential functions, we should make two points. First, there is no expectation that a suitably derived EPP should necessarily resemble the real pair term in the complete non-additive potential. Thus, comparisons of true dimer potentials and EPPs are somewhat artificial, and this should be borne in mind during later discussions. Secondly, equation (7) illustrates that an EPP can vary with temperature, and also density. Hence, strictly, an EPP should be applied only to the assembly for which it is designed: any extension to a different temperature or density assembly implicitly assumes this dependence on the changed variable is weak, an assumption which has not been justified. Thus, application of an EPP designed for water at 25 °C and 1 atm pressure would *not* be expected to give a good second virial coefficient (essentially a dimer property), or to give good results in heterogenous systems such as interfaces where local density may vary. Conversely, a true dimer potential derived from a knowledge of dimer properties will be unlikely to give good results in condensed phases; the achievement of good results must imply *either* the three- and higher-body effects are structurally negligible, *or* the existence of some compensating factor. Unless it can be shown that three-body effects are so well behaved that they result in the same kind of structural perturbations in all aqueous systems of interest, or that such perturbations are acceptably small, then apart from considering limited ranges of temperature and density of homogeneous liquid water, the weight of evidence suggests the only physically acceptable way to move forward in general aqueous systems is through the specific consideration of three-body effects. We might note in passing the strong many-body effects in liquid metals; EPPs for those systems are necessarily density-dependent. The same is true in principle of aqueous systems, though the density dependence may be weaker.

4.2. *Potential Function 'Design'*

Consider two water molecules; their relative position and orientation is defined in terms of the vectors $(\mathbf{X}_1, \mathbf{X}_2)$ which are functions of the Cartesian and orientational vectors $(\mathbf{R}_1, \mathbf{\Omega}_1)$ and $(\mathbf{R}_2, \mathbf{\Omega}_2)$. In principle, each molecule requires three position and three orientation coordinates; the *relative* configurations of two molecules can, however, be fully described by a six-dimensional vector, the reference molecule being placed at the coordinate system origin in a standard orientation.

Thus, to define the water pair potential function, be it a 'true dimer' (TD) or 'effective pair' potential (EPP), we must both *devise* and *parametrize* a function of one distance and five angular variables. This is far from being an easy operation. Ideally, to both devise and parametrize the functional form, we require data on energy as a function of all six variables, and such is certainly not available experimentally except in very limited regions of the two-molecule configurational space. In practice, the choice of the form is usually guided by a mixture of intuitive chemical and structural arguments, supplemented by any experimental data we may have on condensed phase structure (e.g. ices). Although such arguments (e.g. those involving sp^3 hybridization and tetrahedral ice structures which lead to a form which emphasizes tetrahedral orientational dependence) may appear reasonable, they remain intuitive and subject to revision in the light of improved knowledge.

The data used to parametrize the chosen form can be either experimental or theoretical. Bearing in mind the difficulties inherent in using true dimer potentials for condensed phase calculations (the problems of many-body effects), we would prefer to parametrize EPPs on experimental data, in order to compensate as far as possible for the assumption of pairwise additivity. Extensive experimental structural, spectroscopic and thermodynamic data do exist; however, they will generally be restricted to regions of configurational space close to minimum energy configurations. Thus, a function parametrized on experimental data may be seriously in error for configurations away from those at which it was parametrized. Provided the functional form chosen is good, this should not raise too much of a problem; however, a sceptic might (and perhaps a scientist should) argue that we really have no *a priori* way of knowing how 'good' a representation the chosen functional form is. In practice, we will be left with an essentially unquantifiable uncertainty.

In principle, we can use quantum mechanical methods to calculate, to a chosen level of approximation, the potential energy of any dimer configuration, and so ensure a sensible distribution of sampling points to which to fit our chosen form. We could even use such calculations, again in principle, to obtain full information on the topography of our six-dimensional energy surface, and so guide sensibly our choice of form. In practice, however, computational restrictions prevent the generation of a dimer energy database that is both sufficiently accurate and sufficiently extensive; we are normally limited to sampling a relatively small number of configurations, preferably chosen sensibly with respect to the functional form to be fitted. Even here, we are severely restricted computationally, sufficiently accurate calculations requiring large basis sets and the inclusion of electron correlation, without which the important dispersion attraction is entirely absent. There are also technical problems concerned with basis-set superposition errors. Taken together, with respect to typical thermal energies of the order of kT, there may be serious problems inherent in such quantum-mechanical supermolecule calculations for the accurate definition of dimer potentials; recent work to this effect is discussed later. We note at this point, however, that such calculations aim to obtain true dimer potentials, rather than EPPs. Unless some device is included to handle the effects of non-additivity (such as dipole polarizability) we would expect such potentials to have relatively limited success in reproducing condensed phase structure and thermodynamics. However, we should note that quantum-mechanical calculations using all but the very largest and highest quality basis sets generally overestimate the molecular dipole moments of water by some 10–20 per cent. Noting that one effect of non-pair-additive effects appears to be to enhance the effective dipole moment over the isolated molecule value, this defect may coincidentally produce potentials that are more successful than we might otherwise expect. A 'truer dimer' potential derived from limited basis-set calculations may thus turn out to be a reasonably effective EPP.

4.3. *Common Functional Forms*

We list here some aspects of functional forms within which most potentials have to date been cast. Of particular interest is the general similarity of the approaches, with a few exceptions. The nature of the functional form can usefully be considered in terms of standard contributions to the potential from physically identifiable effects, namely overlap repulsion, electrostatic attraction, dispersion attraction and induction. Indeed, there is a strong case for approaching the problem by tackling each of these contributions separately, as in quantum mechanical perturbation theory[**41**] fitting physically sensible functional forms to each of these terms individually.

Overlap repulsion has generally assumed an inverse twelfth-power dependence, largely on the basis of the frequent use of this term in the Lennard–Jones 6:12 potential. However, the theoretical justification for this form is weak: it seems to have been chosen largely for computational convenience ($r^{12} = (r^6)^2$) in pre-computer days. An exponential function has stronger theoretical justification, though few water potentials use it. Normally, a single-centred repulsive term is used, either scaled to isoelectric neon, or obtained by adjustable parameter fitting to experimental or theoretical data. There are limited examples of three-centre repulsions (normally centred on oxygen and hydrogen nuclei respectively). All these functional forms seem to relate to intuitive ideas on the nature of the electron distributions in the water molecule, prejudiced by simple-liquid work. There would appear to be a strong case for examining these assumptions critically. Brobjer and Murrell[**41**] have proposed a bipolar exponential form for HF of the form

$$E_{\text{rep}} = \exp[A(\boldsymbol{a}) + B(\boldsymbol{a})R], \tag{8}$$

where $\boldsymbol{a}$ are angular coordinates. Even for this simpler molecule, reasonable representation of the perturbation theory results required expansions of A and B to six and five terms respectively. There is a strong implication here that our simple repulsive models for water are probably inadequate.

For polar molecules, the *dispersion* term would appear less important than for non-polar molecules, but as we shall see, it is necessary to obtain reasonable representations. Most (though not all) proposed water potentials are limited to the leading r^{-6} term of the more general expansion in which terms in r^{-8} and r^{-10} occur, possibly together with an exponential damping term. The C_6 coefficient of r^{-6} has usually been obtained by either assuming the isoelectronic neon value, or treating it as a fitted parameter.

The *electrostatic* term dominates the medium-range part of the potential, and it seems generally assumed that the orientational dependence of this term controls the directionality of water–water interactions. Functional forms are based on *either* intuition on where the charge centres are thought to be (e.g. on the protons, and lone-pair 'positions' in Bjerrum-type four-point charge models), *or* fitting the quantum mechanically derived charge distributions to a point-charge distribution or multipole moment expansion. There is considerable argument in this area as to the best way of representing the charge distribution. A limited number of point charges is simple for computation, but this limited number may be inadequately flexible. The multipole expansion is appealingly 'clean', but a single-centre expansion has severe convergence problems[**32**]; these appear to be largely overcome by the distributed multipole analysis of Stone and coworkers[**42**], in which the earlier multipole convergence at least partly compensates for the additional number of multipole *centres* required over single-centre expansions. We suspect that largely for computational reasons (which may or may not be valid), most effort to date has been using three- or four-point charge models, usually placed on or near expected charge centres.

Induction effects have largely been ignored, though where they have been included in non-pair-additive models, they have been restricted to dipole polarizability, usually assumed to be linear. An *isotropic* polarizability seems a good approximation, theoretical calculations showing the anisotropy is small ($\lesssim$ 5 per cent), and simulation calculations with larger anisotropies than this showing no detectable effect on ensemble average properties.[**43**] The work on HF[**41**] suggests there is a case for a more detailed assessment of this term.

All models have assumed the water–water interaction can be modelled without considering *charge transfer*, i.e. ignoring any covalent contribution to hydrogen bonding. On the basis of extensive quantum chemical calculations, this seems to be generally accepted as justified.

With the above background, we now summarize a representative sample of dimer potentials, both of 'true dimer' and EPP types.

4.4. *Proposed Potential Functions*

4.4.1. *Pair Additive Rigid Body*. (*a*) *Bernal–Fowler*.[**6**] Being the first quantitative water potential to be proposed, this model is of particular historical interest. Considering how closely the resulting potential resembles many modern potentials, yet remembering the data and theoretical tools available at the time, its derivation was a *tour de force*. The structure of ice was still being argued, London's dispersion theory was only three years old [**44**], the geometry of the water molecule was only just beginning to be spectroscopically defined [**45**], and theoretical chemistry was in its infancy. There were no computers, thus enabling most of the work to be done in one night at a fog-bound Moscow airport.

Bernal and Fowler knew that water was tetrahedrally coordinated in ice-Ih; yet this did *not* lead them to assert that the water–water potential itself should be tetrahedral. They preferred instead to use their chemical intuition, spectroscopic information, and the known intermolecular separation in ice. Their first approximation put full electronic charges on the two proton mass centres, and a charge of $-2e$ on the oxygen mass centre. Because this gave too high a dipole moment (three times the experimental value), they argued first that there must be charge screening, and secondly that the centre of negative charge must be shifted along the HOH bisector in the direction of the positive charge (figure 7). With hindsight, this seems a remarkable thing to do; fitting point charge models to modern quantum-mechanical calculations produces the same effect, although ideas of lone-pair electrons would imply the negative charge centre might be shifted to the opposite side of the oxygen centre, *away* from the hydrogens.

Allowing the assigned charge q and negative charge centre position x (figure 7) to vary subject to the overall dipole moment (incidentally taken to be 7 per cent higher than the known isolated molecule value to try to allow for non-pair-additive effects!) resulted in restricting $0 \leqq x \leqq 0.37$ Å. There being inadequate data to fix both parameters, they argued that the oxygen core electrons would lead to the centre of charge being towards the oxygen nucleus. The final parameters of $q = 0.49e$ and $x = 0.15$ Å might interestingly be compared with the recent TIPS2 (transferable intermolecular potential functions) parametrization of the same basic model ($0.535e$ and 0.15 Å respectively). The dispersion form was obtained from London theory, using the known values for polarizability and ionization potential to obtain the constant for the r^{-6} coefficient. Repulsion was assumed to go as r^{-12}, the

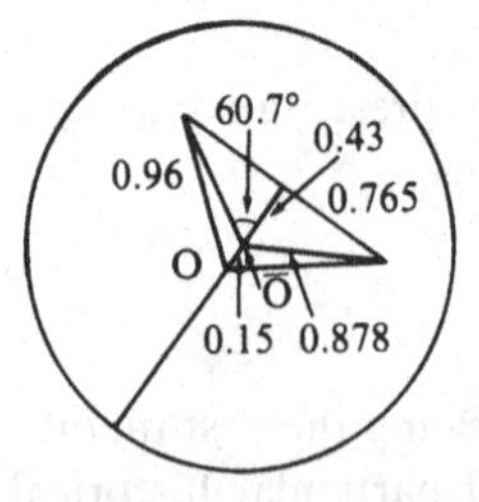

Figure 7. The Bernal–Fowler water model. The centre of negative charge is moved along the symmetry axis in the direction of the positive charge from O to Ō. In the original model, Ō is also the centre of the repulsion and dispersion terms, though in the calculations quoted, the repulsion/dispersion centre has been left at O.

Table 2. *Three-point-charge model parameters (see figure 11)*

Model	r_{OH}(Å)	r_{OM}(Å)	$q(e)$	θ(°)	A[a](kcalÅ^{12} mol^{-1})	$-C$[b](kcalÅ^{6} mol^{-1})
BF[c] original*	0.8776	0	0.49	121.3	0.5604	837.0
modified†	0.96	0.15	0.49	105.7	0.5604	837.0
SPC*	1.0	0	0.41	109.47	0.6294	625.5
TIPS*	0.945	0	0.40	104.52	0.580	525.0
TIPS2†	0.9572	0.15	0.535	104.52	0.6950	600.0
TIP3P*	0.9572	0	0.417	104.52	0.5820	595.0
TIP4P†	0.9572	0.15	0.52	104[52	0.6000	610.0
Watts*	0.96	0	0.32946	104.5	d	625.45[e]
RWK1†	0.9572	0.26	0.60	104.5	d	625.45[e]
RWK2†	0.9572	0.26	0.60	104.5	d	625.45[e]

* Three-point-charge, three-centre models.
† Three-point-charge, four-centre models.
[a] Overlap repulsion coefficient in r^{-12}.
[b] Dispersion attraction coefficient in r^{-6}.
[c] The 'original' BF model [6] differs from that used by Reimers *et al.* [**61**] modified in placing the Lennard–Jones centre on the negative charge centre. The original parameters given here are therefore within the form of the three-point-charge three-centre models.
[d] See text: repulsion is of exponential and Morse forms, and the eight parameters are given in reference 61.
[e] In addition there are C_8 and C_{10} terms which differ between the three potentials (see reference 61).

coefficient being fitted by setting the nearest-neighbour force to zero at the ice-Ih separation distance. Interestingly, Bernal and Fowler also considered the possibility of using an inverse exponential form for the repulsion. Unlike the version of the model used by Reimers *et al.* [**61**], where the oxygen centre O is used, both repulsion and dispersion were centred on $\bar{O}$ (figure 7). The energy surfaces and simulations discussed below refer to the Reimers *et al.* (modified) version, which is probably an improvement over the original formulation.

The full parameters of both versions are given in table 2. We note the strong similarity of the modified version to the present four-site three-point-charge class of models, with even similar parameters. The original model contains what would be considered even by present standards reasonable models of the electrostatic, dispersion and repulsive terms, and the enhanced dipole moment of 2.0 Debye shows an appreciation of the importance of induction, even if the enhancement is probably too small. It produced an ice sublimation energy within 3 per cent of the then accepted value, though Bernal and Fowler pointed out the agreement *must* have been fortuitous, and that the result obtained was very sensitive to small modifications to both the chosen forms (e.g. r^{-12} or exponential repulsion), and model parameters (such as the location of the negative charge centre). We are tempted to conclude that, had a computer been available at the time, Bernal's 50-year start might well have put us all out of a job.

(*b*) *Rowlinson*. [**46, 47**] To complete the case for there being little new under the sun as far as water potentials are concerned, we discuss the early work of Rowlinson, who fitted model water potentials to the second virial coefficient and to the lattice energy and structure of ice. Three models were developed, which relate to multipole moment expansions, three-point and four-point charge representations of the molecular charge distribution; there is also a specific induction term derived from polarizability, so most of the elements of modern water potentials are represented.

In all three models, overlap repulsion and dispersion terms were assumed of inverse twelfth and sixth powers respectively and taken to be spherically symmetrical. Induction was handled through the average induction energy dependent on the mean polarizability; as this is also an r^{-6} term, it was added to the dispersion term in the calculation of the virial coefficient, though it was calculated specifically when examining the ice structure to take account of its orientational dependence. The electrostatic part was dealt with differently in the three models; the first used a simple dipole, which Rowlinson considered inadequate for solid-state calculations. He proceeded to develop a three-point charge model with charges placed on the atomic centres; presumably for reasons of ease of calculation (again this had to be done without computers, and hence analytically), the resultant point dipole and quadrupole moments were calculated (centred on the centre of charge), and these values used in subsequent calculations of the second virial coefficient. The potential could thus be thought of as a Lennard–Jones 6:12 (centred on the centre of charge), plus a three-point charge model designed to give a correct dipole moment; the potential energy of two such interacting models is, however, considered in terms of the first two terms in the multipole expansion (dipole–dipole and dipole–quadrupole). A better representation was thought necessary for solid-phase calculations, so the dipole–octupole and quadrupole–quadrupole terms were included there. The free parameters were fitted to the second virial coefficient.

Further considerations of the ice structure, in which Rowlinson demonstrated the importance of the dipole–octupole term, led to a modification of the original three-point charge model which required an increased quadrupole. Interestingly, this was *not* obtained by moving the negative charge centre towards the hydrogens along the symmetry axis as Bernal had done, such a charge distribution being regarded as less probable than splitting the negative charge into two charges above and below the molecular plane as indicated in figure 8. Rowlinson regarded the precise nature of the O–H bond as irrelevant, as there would be splitting of the charge whether the bond was pure p or an sp^3 hybrid.

As in the Bernal and Fowler work, we note in Rowlinson's potentials

Table 3. *Four-point-charge model parameters (figures 8 and 9)*

	r_+[a](Å)	r_-[a](Å)	θ_+[a]	θ_-[a]	q_+[a](e)	σ[b](Å)	ϵ[b](kcal mol^{-1})
RN	0.96	0.2539	105°	180°	0.3278	2.725	0.7070
BNS (revised)	1.0	1.0	109.47°		0.19562	2.82	0.076472
ST2	1.0	0.8	109.47°		0.2357	3.10	0.075750
STO-3G	0.9572[d]	1.03	109.47°	104.52°[d]	0.1073	—[c]	—[c]

[a] Distance (r) from origin, angle (θ) subtended by, and value of (q), charge (+ or −).

[b] Lennard–Jones 6 : 12 ϵ and σ values.

[c] Both overlap repulsion and non-electrostatic attraction are centred on each oxygen and hydrogen centre, using a 12 : 6 : 3 form. The nine fitted parameters are given in reference 52.

[d] Monomer values specified in the STO-3G dimer calculations.

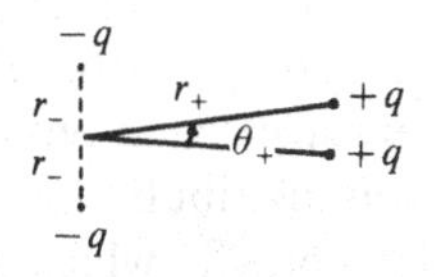

Figure 8. Rowlinson's model in which the quadrupole is fitted by splitting the negative charge above and below the molecular plane at the oxygen centre (cf. the O→Ō movement chosen by Bernal and Fowler in figure 7).

elements of several popular current functional forms. The three-point charge model relates to modern three-point charge models. Rejecting the moving of the negative charge centre *towards* the hydrogens to obtain an improved quadrupole, in favour of splitting the charge as in figure 8, resulted in the first example of a four-point charge model, developed later by Ben-Naim, Stillinger and Rahman. **[48–50]**

Multipole expansion respresentations of these models are also used, though presumably to facilitate calculations rather than because of a preference for this description. We note also the inclusion of an induction energy to account for dipole-induced non-pair-additive interactions. While this is allowed for as an average polarization energy in the second virial calculation, its specific orientation dependence is used in the condensed phase (ice) calculations, which themselves give lattice energies in good agreement with experiment. The parameters of Rowlinson's four-point charge model are given in table 3.

(*c*) *Four-point charge models.* Ice structures are essentially tetrahedral: each molecule is coordinated to four neighbours through hydrogen bonds between the protons of one molecule and the unpaired electron region of its neighbour. The vapour phase HOH angle of 104.5° is sufficiently close to the ideal tetrahedral angle of 109.5° for this tetrahedral organization to be considered reasonable from the proton's viewpoint.

In discussing ice structures, Bjerrum proposed a water model that would be consistent with such a tetrahedral hydrogen-bonding structure. This model placed two positive and two negative charges at the corners of a regular tetrahedron (figure 9), the positive charges relating to the proton, the negative to the lone-pair electron regions. The model is thus charge-symmetrical; it is also in terms of directionality consistent with classical ideas on sp^3 hybridization, though there seems in retrospect little reason for placing the negative charge centres as far from the oxygen centre as the protons. This approach builds into the model for the water–water interaction the precise directionality that is observed in the structure of ices. Although this line of argument may seem reasonable, it involves a confusion between two-body properties (the water–water interaction) and many-body structural effects (the tetrahedrality of ices). Whereas a tetrahedral directionality at the two-body level will very likely result in tetrahedral crystal structures, it is not necessary to postulate potential function tetrahedrality in order to obtain a tetrahedral coordination in a crystal structure. This point should be stressed, as in order to understand water both in the pure liquid and in mixed systems, strong tetrahedrality can be misleading, and in fact quantum-mechanical calculations show clearly that a strong tetrahedral model for the electron distribution is wrong. We should note here in this context (i) the discussions on g_{OO} which explain that the origin of the second, often misnamed tetrahedral, peak shown *not* did require strong specific *tetra*hedrality in the intermolecular geometry, and (ii) no such tetrahedrality assumption was considered necessary by either Bernal and Fowler, or Rowlinson, even though they clearly knew about the tetrahedral nature of the ice-Ih structure. Many-body structures, in both solid and liquid states, are Boltzman ensembles which can depend upon complex interplay between the repulsive and attractive regions of the potential function. To impose *directly* the symmetry of the condensed phase onto the symmetry of the intermolecular potential function may itself lead to an inflexible model. We can use such information for guidance in potential

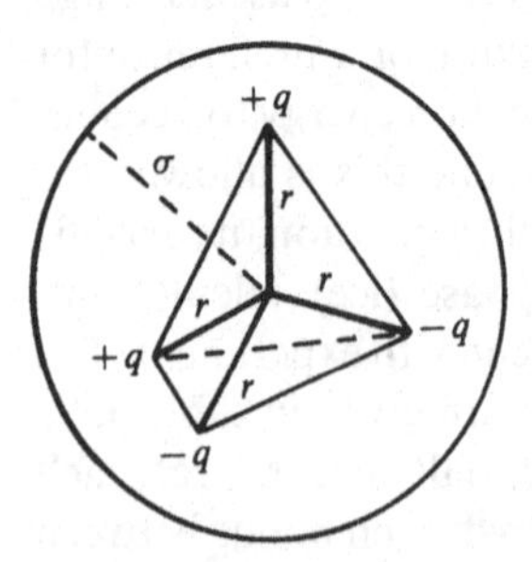

Figure 9. The classical Bjerrum four-point-charge model, in which charges of $\pm q$ are placed at the vertices of a regular tetrahedron. Reproduced, with permission, from reference 49.

function choice, but should avoid assuming that solid-state geometry itself can give directly the complete orientational dependence of the potential function.

The subsequent history of four-point charge models bears out these points. The initial model used in molecular dynamics calculations by Rahman and Stillinger[**48**] was that devised as an EPP by Ben-Naim and Stillinger.[**49**] The initial parametrization of this BNS model took a Lennard–Jones 6:12 form for the (assumed spherically symmetric) dispersion and repulsion terms (as Rowlinson), and chose accepted neon values on the basis of its being isoelectronic with water. The symmetrical Bjerrum model (figure 9) was used for the electrostatic part, with $r_{OH} = r_{OLP} = 1.0$ Å. To avoid non-physical close approaches between charge centres on neighbouring molecules, a switching function which vanished at sufficiently small distances was invoked. Both the second virial coefficient and equilibrium O···O distance in hexagonal ice were used to fit the charge and the switching function parameters. Using *both* a true dimer property *and* condensed phase (ice) data might seem in hindsight a little contradictory if an *effective* pair potential is the aim. Interestingly, the fitted charge parameter gave a dipole moment some 10 per cent greater than the isolated molecule value; there is thus some account taken, through the enhanced dipole moment, of polarizability effects. As a result of initial molecular dynamics calculations, the original function was rescaled, increasing both the charge (and hence dipole moment) and the Lennard–Jones ϵ parameter. The renormalized parameters are given in table 3.

Further modifications to the model parameters were made on the basis of the simulated BNS liquid being too tetrahedral, plus suggestions that the vibrational frequencies would be too high.[**50**, **51**] The major change was to bring in the negative charge centres by 0.2 Å, thus breaking the charge symmetry. This involved readjustment of the other model parameters; any identification of the Lennard–Jones parameters with those of neon was abandoned, and no attempt was made to fit the second virial coefficient, thus removing the apparent inconsistency in using a dimer property to parametrize an EPP. The dipole moment is further increased over BNS to 2.35D, some 27 per cent higher than the isolated molecule value, and therefore implicitly including a greater degree of dipole moment enhancement consequent upon inductive effects. The parameters for ST2, one of the most widely used water models, and one of the few purposely EPPs, are given in table 3.

Jorgensen later fitted a four-point-charge model to STO–3G calculations of the energies of 266 (later 291) water dimers[**52**], for which a satisfactory three-point-charge model fit could not be obtained. The basic tetrahedral geometry was that of ST2, with adjustable lone-pair positions. Unlike ST2, however, this form included Lennard–Jones 6:12 dispersion and repulsion terms on each of the H and O atoms; the dispersion and repulsion are therefore anisotropic, presumably removing the need for a switching function.

An additional atom–atom r^{-3} term is argued to account for dipole–dipole interactions (constrained to vanish at larger r). The fitted parameters are given in table 3. Interestingly, the charge values are low, resulting in a dipole moment of $1.22D$, lower than the isolated molecule value. Jorgensen argues that this is compensated for by the r^{-3} term included in the form.

(*d*) *More than four-point charges.* Four-point charge potentials model the positive region in terms of shielded protons, and the negative region by dipole-moment consistent point charges. This form can be extended by representing *all* electrons and nuclei by suitably placed point charges, which in the case of water are indicated in figure 10. The form EPEN/2 (empirical potential based on interactions of electrons and nuclei) is based on indications from the monomer molecular orbitals; the adjustable parameters (repulsion, dispersion and charge centre positions) are fitted from experimental and/or theoretical data. One argument advanced in favour of this relatively complicated, and computationally expensive form is the identification of the elements of the form with the electrons and nuclei themselves, thus reducing the degree of intuition in the design of the functional form. This should result in greater transferability of the parameters between related molecular groups, and hence provide a framework for calculations on more complex, mixed aqueous systems. Both a dispersion (r^{-6}) and a repulsion term are centred on the molecular centre of mass. The latter term is taken as an inverse exponential in preference to the rather hard r^{-12} form used in the potentials discussed so far. This seems to have been preferred initially for computational convenience (ease of calculating derivatives); as mentioned earlier, the exponential form has also a sounder theoretical justification.

Three parametrizations have been published. The original version of Shipman and Scheraga [53] used experimental ice and vapour phase data to fit 17 parameters; the mixing of condensed and vapour phase data again results in a model that cannot be regarded as either a true dimer or an EPP. The isolated molecule dipole and quadrupole moments were fitted; thus no dipole moment enhancement was included to account for average effects of non-pair additivity. A later modification used the quantum mechanical CI

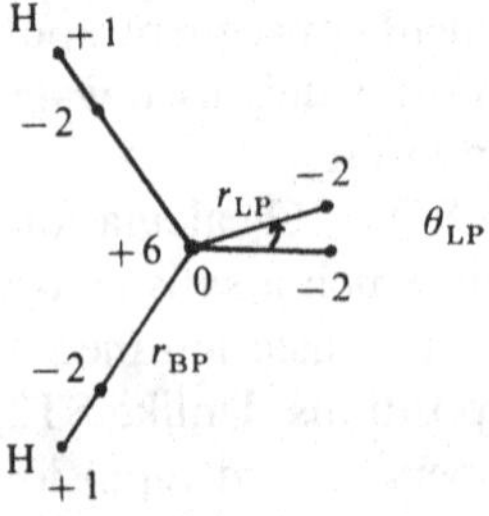

Figure 10. Sketch of the EPEN model, in which charges are positioned with respect to expected charge centres.

energies of 66 dimer configurations calculated by Matsuoka *et al.* [54] in deriving the MCY (Matsuoka, Clementi & Yoshimine) potential discussed in (*e*) below; this version should therefore be regarded as a true dimer potential. The same holds for the recent QPEN parametrization of Marchese *et al.* [55] who used the same 66 dimer energies, together with an additional 229 dimer energies calculated from the final MCY potential itself. The QPEN potential thus might seem to be an alternative representation of the MCY potential, and hence contain no new information. This is not quite correct, as the parametrization used energies calculated from both MCY itself and the original quantum mechanical calculations which we shall see later (section 6.2) are not fully consistent with each other.

(*e*) *Three-point charge models: Hartree Fock* (*HF*) *to Matsuoka, Clementi and Yoshimine* (*MCY*) *and beyond.* A series of potentials has been developed by Clementi and coworkers, on the basis of quantum mechanical calculations on the water dimer. Hence, given (i) an adequate data base from the quantum mechanical calculations, (ii) choice of appropriate form for the potential function and (iii) satisfactory fitting of the parameters describing the form, the result should be a good true dimer potential capable of reproducing dimer data such as the second virial coefficient and equilibrium dimer geometry. Given the significance of non-pair-additive effects, we should not expect to obtain a good effective pair potential for liquid state calculations. The degree to which the above three conditions are met will be discussed subsequently in the light of experimental comparisons and the problem of using quantum mechanical dimer calculations.

Hartree-Fock (*HF*) *potential.* Using a relatively high quality basis set [56, 57], energies corresponding to 190 selected dimer geometries were calculated, using fixed water molecular geometry. An additional 26 points were chosen in the vicinity of the energy minimum, to improve the data base in this region which would clearly be more frequently sampled in simulation calculations. A three-point charge representation similar to that of the modified Bernal and Fowler model (see (*a*) above) was chosen to give an analytical formula that 'best combines numerical accuracy with mathematical simplicity'. For the basic model of figure 11, we can write an expression for the energy of a pair of molecules as

$$\begin{aligned} E = {} & q^2\left[\frac{1}{r_{11'}}+\frac{1}{r_{12'}}+\frac{1}{r_{21'}}+\frac{1}{r_{22'}}\right]+\frac{4q^2}{r_{33'}}-2q^2\left[\frac{1}{r_{13'}}+\frac{1}{r_{23'}}+\frac{1}{r_{31'}}+\frac{1}{r_{32'}}\right] \\ & +a_1\exp(-b_1 r_{00'}) \\ & +a_2[\exp(-b_2 r_{11'})+\exp(-b_2 r_{12'})+\exp(-b_2 r_{21'})+\exp(-b_2 r_{22'})] \\ & +a_3[\exp(-b_3 r_{10'})+\exp(-b_3 r_{20'})+\exp(-b_3 r_{01'})+\exp(-b_3 r_{02'})], \quad (9) \end{aligned}$$

where the prime indicates the equivalent centre on the neighbouring molecule. The first three terms give the electrostatic contributions, while the last three account for repulsion between oxygens, hydrogens and oxygen–hydrogen respectively. We note that repulsion is exponential rather than inverse twelfth,

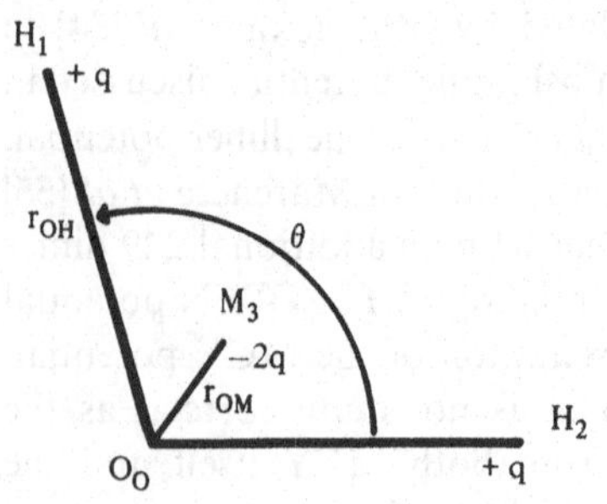

Figure 11. The basis three-point-charge, (four-centre) model.

and that, because there are repulsions centred on each nucleus, the repulsion is anisotropic. No induction terms are included explicitly, though in principle two-body inductive effects will be partly included in the other parameters. There is no provision for a dispersion term: dispersion effects are absent in Hartree-Fock calculations, as no account is taken of electron correlation. We note the similarity of the electrostatic form with the original Bernal–Fowler proposal, including the displacement of the negative charge centre.

Table 4 gives the resulting fitted parameters. The overall standard deviation of the fit is 1.13 kcal mol^{-1}, most of this coming from the repulsive configurations which will be sampled less frequently in a room temperature simulation. For the 80 attractive configurations, the fit is much better, with a standard deviation of less than 0.2 kcal mol^{-1}. Although the form of the potential precludes direct comparisons between the repulsive parameters of other three-point charge models, the charge values and negative charge centre displacements of 0.67e and 0.231 Å are much larger than the 0.49e and 0.15 Å values in the original Bernal and Fowler potential. On the basis of an additional 13 configurations, a revised HF potential was devised.[**58**] The revised parameters are shown in table 4.

HF and dispersion. A serious drawback to the HF potential is its lack of a dispersion term; the attractive part of the potential would thus be expected to be insufficiently deep, and to fall off too slowly with distance. In attempting to correct for this, Kistenmacher *et al.*[**59**] added dispersion terms from different sources. They considered both the induced dipole-induced dipole (London) term that would be expected to dominate at large r, and also other contributions that would be significant at shorter distances.

A semi-empirical form for the correlation energy E_{corr} suggested by Wigner[**58**] was used to estimate E_{corr} for a variety of orientations and configurations. Finding E_{corr} relatively intensive to orientation, the results were fitted to an exponential form for a short-range correction term

$$\Delta E_s = -Ce^{-\delta r_{OO}} \qquad (10)$$

where r_{OO} is the O–O separation distance. At the minimum separation, this results in a deepening of the potential well of this HFW potential by 0.25 kcal

Table 4. *Clementi-type potential parameters*[a]

a	HF	Revised HF	MCY/CI	MCY/inter (MCY)	Bounds	CH	CC
$q(e)$	0.670363	0.647632	0.751743	0.717484	0.717484	0.628217	0.65789
r_{OM}(a.u.)	0.435	0.427	0.487741	0.505783	0.505783	0.471636	0.47231
a_1	582.277054	113.996966	1864.271482	1734.196000	1663.238	72.317219	724.1307
a_2	0.143789	1.242839	0.662712	1.061887	0.564	1.603418	5.7051
a_3	5.470184	6.508362	2.684452	2.319395	6.636	0.0079698	3.7303
a_4	—	—	0.675342	0.436006	4.915	9.363718	0.7297
a_5	—	—	—	—	—	0.0079520	—
b_1	2.520593	2.100760	2.753110	2.726696	2.6627	1.892938	2.5165
b_2	1.221756	1.653882	1.299982	1.460975	1.2698	1.896610	2.0345
b_3	1.936626	2.071387	1.439787	1.567367	1.2921	0.585797	1.6808
b_4	—	—	1.141494	1.181792	1.2190	2.347936	1.3127
b_5	—	—	—	—	—	0.373522	—

[a] Parameters are defined in figure 11, eqns 9 and 12. a_5 and b_5 are additional H–H repulsion parameters explained in the text. Units are atomic units of energy (a_i) and distance $^{-1}(b_i)$. Values are given to the number of significant figures specified in the original papers, except that the CH values have been truncated.

mol^{-1} (although the S and C parameters quoted in [59] seem inconsistent with the results in figure 3 of that paper). For the longer range dispersion term E_D, the leading term of the standard form is used:

$$E_D(r) = -C_6 r^{-6}. \tag{11}$$

Two extreme values for C_6 are used, derived from the Kirkwood-Müller [60], and London relations. [60] As the polarizability tensor of water is not significantly anisotropic, no orientational dependence of E_D is considered. These HFL and HFK modifications lower the energy minimum by about 1.0 and 2.0 kcal mol^{-1} respectively. We should note that addition of the dispersion term shifts the potential energy minimum to slightly lower distances (0.02 to 0.1 Å), as would be expected. Figure 12 shows these effects for HFK and HFL. A final set of modified potentials (HFW(s) and HFK(s)) were constructed by adding the Wigner short range correction (eqn (10)) to the HFW and HFK dispersion-corrected models. As the short-range (W) and long-range (K or L) terms do not match at short range, the long-range term is effectively switched off at $r \leqq 3$ Å. The mismatch between the short-range and long-range terms at this separation presumably means that the long-range term is modified in some continuous r-dependent fashion. An alternative dispersion correction has been proposed by Reimers *et al.* [61], which uses r^{-6}, r^{-8} and r^{-10} coefficients derived from work by Scoles *et al.* [62–64]

Matsuoka, Clementi and Yoshimine (*MCY*). To obtain the dispersion term from quantum mechanical dimer calculations necessitates going beyond Hartree-Fock self-consistent field methods to include the effects of correlation energy. An attempt to do this was made by Matsuoka *et al.* [54] using configuration interaction (CI) calculations to produce dimer energies with

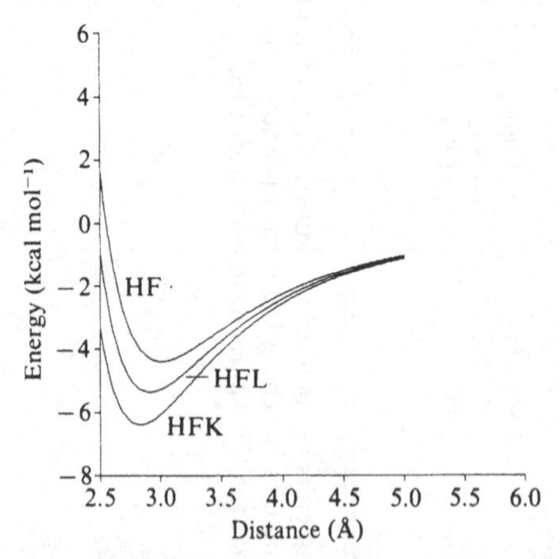

Figure 12. The effect of the London (HFL) and Kirkwood-Müller (HFK) dispersion corrections to the distance dependence of the HF potential. C_6 terms taken from reference 60. Standard configuration ($\theta_D = 52.26°$, $\theta_A = 120°$).

which to reparametrize the HF form (eqn (9)) modified by the addition of a dispersion attraction term

$$-a_4[\exp(-b_4 r_{10'})+\exp(-b_4 r_{20'})+\exp(-b_4 r_{01'})+\exp(-b_4 r_{02'})]. \qquad (12)$$

We note from the form of eqn (12), referring to figure 11, that the dispersion term is calculated in the final potential only with respect to the O–H intermolecular separation. This implies it will have an orientational dependence.

As we shall mention subsequently, the technicalities of CI calculations are not straightforward, there being considerable uncertainty as to how best to handle intramolecular correlation energy, basis-set superposition errors and size-consistency corrections. Matsuoka *et al.*'s approach to the problem is described in reference[**54**], where they calculate dimer energies of 66 configurations. Two sets of parameters of eqns (9) and (12) are fitted, yielding two different potentials labelled CI and 'inter' in the original paper. The former refers to the full CI calculation, which will include *intra*molecular correlation effects, while the latter relates to the SCF (self consistent field) result, corrected for *inter*molecular correlation effects only. The two potentials are significantly different, giving dimer energy differences of at least 0.5 kcal mol^{-1} over a wide range of the configurational space, and minimum dimer separation distances which differ by nearly 0.1 Å, illustrating some of the problems inherent in quantum mechanical dimer calculations, to which we shall return.

The MCY-CI and MCY-inter ('MCY') parameters are given in table 4. We note specifically an increase compared to the HF values in both the charge value and the oxygen-negative charge distance.

Bounds. At the beginning of this section, we specified three conditions relating to the likely success of a quantum mechanically-derived potential function. The third condition was the goodness of fit to the chosen form. The satisfaction of the latter point will depend on the sophistication of the fitting procedure used, and the weighting of the various sample points in the fitting calculations. Recently, Bounds[**65**] pointed out specific features which characterized the deviation between the original CI energies and those calculated from the fitted MCY form, and re-optimized the parameters in the original MCY potential. His new values, given in table 4, give a significantly smaller r.m.s. deviation between data points and the final fitted form than does the original MCY fit, namely 0.10 kcal mol^{-1} (cf 0.62 kcal mol^{-1}) for all 66 points, and 0.06 kcal mol^{-1} (cf. 0.17 kcal mol^{-1}) for the 57 attractive points. As we shall discuss later, this reparametrization qualitatively changes the results of condensed phase simulations, underlining the high sensitivity of simulated ensemble average structure to potential function detail.

Habitz and Carravetta: CH and CC. Two further modifications have been proposed by Clementi and coworkers.[**66**, **67**] Both are further attempts to refine the correlation energy contribution. Using the direct CI (configuration interaction) method, including all single and double excitations, the energies of 169 selected dimer geometries were calculated by Clementi and Habitz. [**66**] A more flexible basis set also was chosen, to try to improve the calculation of the dipole moment. The resulting (CI) dipole moment of $1.87D$ compares very well with the experimental value of $1.855D$, though the consequent lack of any significant dipole moment enhancement would be expected to make this a worse EPP for condensed phase calculations. Two modifications were made to the MCY analytical form. First, a second exponential for the hydrogen–hydrogen interaction was found necessary to improve the representation of the repulsive part of the potential. Secondly, for numerical stability, the attractive (dispersion) O-H interaction was multiplied by the relevant hydrogen–oxygen distance. The additional two parameters required to describe the H–H repulsion are given in table 4 as a_5 and b_5. In the fitting of the data to the form, the energies near the minima were weighted heavily.

A final modification of the MCY form [**67**] results from a calculation of the correlation energy using an approximate expression suggested by Colle and Salvetti [**68**, **69**] which is added as a correction term to an SCF calculation. The 105 dimer configurations used included the original 66 used for the MCY calculation, the remainder being close to the equilibrium geometry. Table 4 lists the final parameters, again fitted to the MCY form. We might note in particular the change in q, which brings it closer to the earlier HF value.

(*f*) *Other three-point charge models.* As are the MCY-type potentials, these are also based on three-point charge models. The SPC (simple point charge) effective pair potential developed by Berendsen's group was designed specifically for water at room temperature and a density of 1 g cm^{-3} [**25**], with the aim of obtaining a model that was computationally simple for use in biomolecule-water molecular dynamics calculations. This pointed to the advantage of placing the point charges at the atomic mass positions to avoid redistribution of forces and torques. The repulsion and dispersion terms are represented by a Lennard–Jones 6:12 form on the oxygen centre, the dispersion coefficient being fixed at the value derived from the London expression; both terms are therefore spherically-symmetrical. Water molecule geometry is fixed at ideally tetrahedral (rather than the experimental HOH angle of 104.5°), with the O–H intramolecular distance set at 1.00 Å. The electrostatic term is modelled by three-point charges on the hydrogen and oxygen nuclei. The hydrogen charge q and repulsive coefficient B remain to be fitted; these are done with reference to the liquid configurational energy and pressure at 300 K, which are obtained from simulation calculations for variations in q and B. In addition, $g_{OO}(r)$ is monitored to ensure that the final fitted values yield the characteristic second peak at about 4.5 Å (figure 1). Interestingly, only in a restricted region of the (q, B) space could water-like PCFs be obtained, and this region showed only small overlap with the target energy and pressure points; this again seems to stress the sensitivity of the ensemble average structure to small changes in potential function parameters. Table 2 gives the final parameters, which were actually fitted through a series of simulation calculations of the state point in question; the SPC model therefore can truly be considered to be an effective pair potential for that state.

Most other three-point charge models originate from Jorgensen's group, whose TIPS (transferable intermolecular potential functions) series[**70**] was designed to give simple, effective pair potentials for fluid simulations, yet which had adequate transferability to allow the successful simulation of a range of organic liquids and solutions. For water, the experimental monomer geometry ($r_{OH} = 0.9572$ Å; $\widehat{HOH} = 104.52°$) is used (cf. SPC, where a slightly different geometry is used, but where otherwise the form of the electrostatics is identical), and a Lennard–Jones 6:12 centred on the oxygen only. The repulsive, dispersion and charge parameters were fitted by examining short ($\sim$ 100 k configurations) Monte Carlo runs using different parameters. Table 2 gives the final parameters, which not surprisingly, considering the similarity in form, are similar to those for SPC. The dipole moment of 2.25D shows a 20 per cent enhancement of the monomer value, again illustrating the need for an EPP to reflect the effects of non-pair additivity on the leading dipole term.

The development to TIPS2 abandoned the simple three-site model, reverting to the modified Bernal–Fowler form (figure 7), with the repulsion and dispersion terms centred on the oxygen. The form is similar to MCY (figure 11), although unlike MCY and its relatives, the hydrogens contribute only to the electrostatic part of the interaction.

Charge, repulsion and dispersion parameters are fitted to give a compromise between correct liquid energy and density, and a water-like g_{OO}. Jorgensen points out the difficulty of obtaining a set of parameters in full agreement with all these constraints. Final parameters are given in table 2. As in TIPS, the dipole moment (2.24D) is enhanced some 20 per cent above the monomer value.

Two further parametrizations have been published by Jorgensen, and the parameters are also reported in table 2. TIP3P is a reparametrization of the three-site TIPS model to give improved energy and density values for liquid water; the problem of fitting both density and the g_{OO} second peak is again stressed by Jorgensen, who states that this peak tends to vanish as the density is improved for the three-site model. TIP4P is an alternative parametrization of the four-site three-point charge TIPS2 which gives a slightly higher density for the liquid.

An interesting set of essentially three-point charge potentials based on an original model by Watts[**73**] has received only limited attention. The original Reimers, Watts & Klein (RWK) model [**61**] placed charges on oxygen and hydrogen as in SPC, values being constrained to give the isolated molecule dipole moment. Interestingly, the dispersion contribution (used in conjunction with a switching function) included C_6, C_8 and C_{10} terms, and the repulsion is anisotropic, being of straight exponential form for OO and HH repulsions, and of Morse form for the OH term. Being fitted to gas phase data, we would regard this as a true dimer potential. Two further variants have been derived from the Watts model. RWK1 was derived by shifting the negative charge centre along the symmetry axis (as in Bernal and Fowler, MCY and TIPS models, figures 7 and 11) to fit additionally the quadrupole moment, thus recognizing the importance of the quadrupole in liquid state structure (recall the discussion of section 3.6). RWK2 was obtained from RWK1 by changing the dispersion terms to what were regarded as more satisfactory values. This necessitated a reparametrization of the repulsive constants, which was accomplished using the second virial coefficient together with the static lattice energies and bulk moduli of three ice phases. It is thus neither an EPP nor a true dimer potential. The parameters of all three models are given in table 2.

4.4.2. *Pair-Additive Non-Rigid Body*. Despite the questionable assumptions inherent in using rigid-body models, only limited attempts have been made to devise non-rigid-body potentials. The central force (CF) model of Lemberg and Stillinger[**74**] was designed to allow intramolecular motions, and possible dissociation, by treating the hydrogens and oxygen separately, with a strong interaction between them designed to retain the molecular geometry. Effective point charges reside on the three centres, and the behaviour of an assembly of particles is thus determined solely by centrally-acting forces. Part of the motivation seems to have been theoretical, in that the water problem might be more approachable theoretically if it could be considered in terms of isotropic potentials. The existence of intramolecular vibrational modes also permits an approach to quantum corrections.

Careful thought has to go into the design of the three potentials v_{OO}, v_{OH} and v_{HH}. The forced simplicity makes the problems of designing these functions particularly difficult, as the interaction of, for example, a proton with a neighbouring (hydrogen bonding) oxygen is very different from that with its covalently bonded partner. This in turn may give inadequate flexibility to handle the subtleties of the orientational dependence of certainly the electrostatic interaction, and possibly the repulsion and dispersion terms. An example is the fitting of the charge to reproduce the correct dipole moment; Bernal and Fowler, and Rowlinson have already demonstrated the need to move the negative charge from the oxygen centre if a reasonable quadrupole moment is to be obtained, yet this is not possible within the restrictions of the central force model. Thus, if the quadrupole is important to the orientational structure of the liquid (as implied by the work of Patey and Valleau[**27**], section 3.6), we have a problem within this model unless somehow we can include at least the quadrupolar effects in the chosen distance dependencies of the OO, OH and HH interactions.

The OH potential necessarily has a deep attractive well with an absolute minimum at the OH intramolecular distance, rising rapidly at lower r to include both the O$\cdots$H attractive electrostatic interaction, but compensated by a second repulsive term. The HH interaction has to be such as to retain the non-linear molecular geometry, and to achieve this using central forces alone requires a local minimum at the intramolecular H–H distance, giving a somewhat odd form to v_{HH}. The various considerations are discussed in detail in reference 74. The forms of the potential (modified subsequently), designed to give correct normal modes of vibration, are shown in figure 13; the various sets of parameters are given in references 74 and 75. The variation possible in r_{OH} gives the model a degree of non-pair-additivity.

One particular problem of such models brought out by the ordinate values of figure 13 is that the energy values given by each $v_{ij}(r)$ are relatively large numbers at typical intermolecular distances (~ 3 Å); thus, for a typical pair interaction, the dimer energy (~ 5 kcal mol^{-1}) will be obtained from the differences between several relatively large energies (v_{OO} (3.0) for example is +45.8 kcal mol^{-1}). Yet many of the parameters of the model are fitted to *intra*molecular data. Although sensible fitting can minimize these problems, any necessary subtleties in the *inter*molecular part of the potential may be very difficult to handle when small *intra*molecular changes can give rise to large energy differences.

In order to try to remedy deficiences revealed in molecular dynamics calculations using CF, revised parameters have been proposed.[**75**] The smallness of these changes in the individual $v_{ij}(r)$, and the significant changes that result in calculated properties, underline the comments on sensitivity above. The significance of the constraints in the model are also noted by Rahman *et al.* [**75**], who point out the problems of improving defects in the model without compromising other constraints which are considered essential.

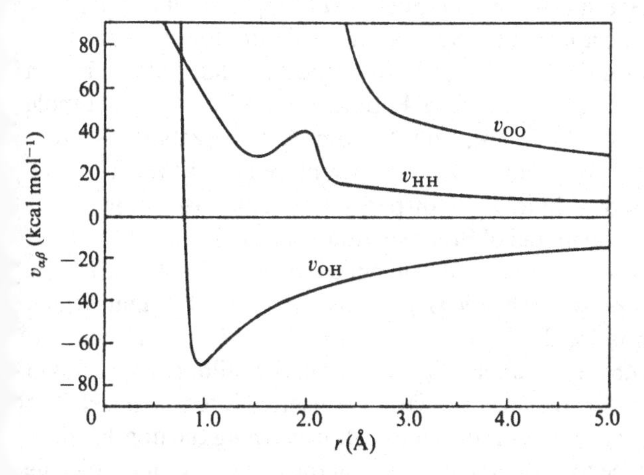

Figure 13. Interaction potentials between oxygen and hydrogen centres of the original CF model. Reproduced, with permission, from reference 74 and 75.

4.4.3. *Non-Pair Additive Models.* All attempts to include non-pair additivity explicitly at the molecular level have used the device of dipole polarizability. In essence, the field at any given molecule gives rise to an induced dipole moment, leading to an enhanced molecular dipole, as is observed experimentally in ice structures. This device uses merely the first of a series of polarizability terms which should all be included in principle. Objections have been raised to the use of polarizability [**76**] but the apparently reasonable results obtainable from what limited tests have been made, coupled with quantum mechanical results showing induction to be the dominant non-pair-additive effect, have encouraged its exploration.

Polarizability can be superimposed on a chosen basic form. Berendsen's early work [**35**] used polarizability in conjunction with ST2, while Barnes *et al.* [**28, 29, 32, 34**] preferred a multipole expansion representation of the charge distribution, allowing only the dipole moment to change under the influence of the field experienced. This so-called polarizable electropole (PE) model can be applied at any desired level of the multipole expansion, the first version PE(QQ) truncating the expansion prematurely after the quadrupole term. [**32**] The experimental dipole moment value is used, with theoretical quadrupoles and higher terms taken from quantum mechanical monomer calculations. Repulsion is handled through an r^{-n} dependence, while the dispersion term is limited to r^{-6}; the parameters were fixed initially from the equilibrium dimer geometry and energy, though data from other sources (especially higher level quantum mechanical calculations, second virial coefficient, ice and water energies and structure) have also been fed into later variants. This model, like all non-pair-additive models, is designed to be effective both in the vapour phase and condensed phases. Such models should therefore, be true dimer potentials that are also valid in condensed phases.

The non-pair-additive model of Campbell *et al.* [**33**] is similar. The first order Coulomb energy is approximated by a high order multipole expansion. The electron distribution used for fitting the multipoles is that obtained from the near-Hartree-Fock wave function of Popkie *et al.* [**57**]; thus, the dipole moment will be too high. As in PE, only the dipole polarizability is used. Repulsion is modelled through an r^{-9} or r^{-12} term placed on the nuclei, while only the r^{-6} term is used to handle the dispersion, assumed isotropic.

Finally, the polarization model of Stillinger and David [**36**] can be regarded as an extension of the central force model, in which the water molecule is now considered as two *bare* (not shielded) protons, and an O^{2-} unit which possesses a form of non-local polarizability. As with the Campbell and PE models, a classical, isotropic polarizability is used; in the Stillinger and David model, however, spatial electron delocalization is inserted into the model. The forms of r_{OH} and r_{OO} together with the constants describing the non-locality, are chosen with reference to experimental data on the isolated water molecule (geometry, dipole moment, force constants, and dipole derivatives), vacuum ionization energy of H_2O, geometry and energy of the water dimer, and the geometry and dehydration energy of $H_3O_2^-$. The details of the fitting, and the final form, are lengthy; they are described in reference 36.

4. 5. *Summarizing Remarks*

Most potential function development for water has been within the three-point or four-point-charge, rigid body formalism, first proposed by Bernal and Fowler and by Rowlinson respectively. Within either form, there are variations in charge magnitudes and positioning and also the treatment of dispersion and repulsive terms. Early models favoured an r^{-12} repulsion, though more recent movement seems towards a theoretically more justifiable, and softer, exponential form. Most models assume a spherical repulsion. Dispersion is modelled usually by the first term in the $C_6 r^{-6} + C_8 r^{-8} + \ldots$ expansion, though some work including the higher terms (leading presumably to a faster fall-off in attraction with distance) has been done, e.g. in the Watts model and its derivatives. Despite the basic similarity of several of these models, there are suggestions (e.g. from the apparent need to reparametrize the first model of a series), of high sensitivity of structure and properties of the simulated condensed phases to the potential function details. A case in point is the strong similarity between SPC and TIPS, with only the former able to give a reproducible water-like second peak in g_{OO}. Moreover, the fact that the SPC parameters chosen appear to be on the boundary of a region of (q, B) parameter space that can yield a $1.6r_0$ peak underlines the high sensitivity of this important structural feature to apparently small changes in the potential function. This sensitivity implies that the formulation of a quantitatively satisfactory water potential will be problematical. It also raises worries that some of the effects we have ignored in concentrating on *rigid body*, *effective* pair potentials may, even though small, be significant in tipping the sensitive balance that seems to control liquid water structure.

We now proceed to look in more detail at the similarities and differences between representative dimer potential surfaces. Although we should be aware of the dangers of extrapolating from dimer properties to condensed phase structural features, we may thereby be able to throw some light on potential function – condensed phase structure relationships.

5. Potential Function Energy Surfaces and Sections

5.1. *The Dimer Surface*

At the dimer level, comparisons between potential functions can be made through consideration of the topography of dimer potential surfaces. This facilitates the recognition of common features between different potentials, and of differences which may relate to different structural tendencies in the condensed phase. Any such conclusions concerning condensed phase behaviour must necessarily be tentative; although in principle full ensemble simulations are needed to allow *firm* conclusions to be drawn about potential function–structure relationships, useful clues can be obtained at the dimer level.

We define our dimer coordinate system with respect to the minimum dimer geometry in figure 14. A total of six independent variables are definable, implying we should examine a six-dimensional surface for each potential function. We concentrate here on only a subset of these variables, namely the O–O separation r_{OO}, donor angle θ_D and acceptor angle θ_A. This limitation of considered variables may prevent the recognition of possibly important aspects of the complete energy surface, though limited evidence suggests these three variables should contain the strongest energy dependencies. For example, the $E(\psi)$ section calculated by Scheraga's group [77], using the large basis set used originally by Hankins *et al.* [30] shows the ψ rotation barrier is not high ($\sim 2kT$); thus all ψ orientations would be expected to be reasonably-well populated in a room temperature simulation. That said, however, potential functions which yield significantly different rotational barriers (e.g. raising the ψ barrier to approximately $3kT$) will result in significantly different populations of the ψ orientations, which may be structurally significant. In the absence of more thorough exploration of such variables, a considerable uncertainty about their structural significance remains.

In this discussion of energy surface sections, we will be particularly interested in (i) the shape of the surfaces in the neighbourhood of the dimer minimum as we would expect this region to be explored most frequently in a room temperature assembly, together with (ii) the relative heights of any apparent barriers. We do not discuss specifically equilibrium dimer geometry and the second virial coefficient: these are all implied in the surfaces. A recent discussion of dimer geometry (which most potentials predict reasonably well) and the second virial coefficient (which nearly all potentials predict poorly) is given in reference 61.

5.2. *Quantum Mechanical Reference Data*

In addition to comparing energy surface sections of different potential functions, we would also like to compare the model surfaces with 'reality'. This requires a set of *reference data*. The only possible source of such a reference data set is high level quantum mechanical calculations, and we employ here the results of our *ab initio* calculations [**79**, **80**] using the very large (11s7p1d:6s1p)→[5s4p1d:3s1p] contracted basis set of Diercksen [**78**], with 35 basis functions per water molecule. No larger basis sets have been used for extensive energy surface studies (that of Clementi and Habitz [**66**] is of the same size) and we believe that this basis set gives as accurate a surface as any other obtained to date.

There are, however, significant problems in using quantum mechanical

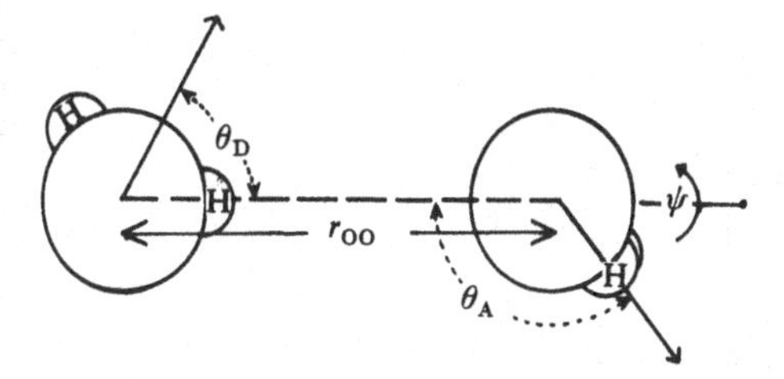

Figure 14. The water dimer coordinate system used. ψ indicates relative rotation about the O–O axis.

reference data. First, all the large basis set calculations agree that the (presumed global) minimum energy structure is the linear *trans* dimer (figure 14), and that the linear *cis* and bifurcated forms are unstable with respect to this. Quantitatively, the experimental dimer geometry is approximately reproduced by the quantum mechanical calculations, though there are possible discrepancies which might be assigned to uncertainties in various corrections that should be applied, namely the basis set superposition error (BSSE) for both SCF and CI calculations, and the size-inconsistency correction of truncated CI calculations.

A detailed examination of the effects of these correlations [**79**, **80**] allows us to conclude that, provided the basis set is adequate, the *shape* of the energy surface can be obtained with confidence. The same cannot be said either for *absolute* energy values or for the relative heights of apparent minima or maxima in a given section. Consider, for example, the donor surface section of figure 15, which shows the dimer energy as a function of the donor angle θ_D (figure 14). The main features of interest are the two apparent minima (corresponding to *trans* and *cis* configurations) and the two barriers between them. Full calculations of this section are given for the SCF calculation without (E_{SCF}) a basis set superposition error correction, and for a CI calculation (E_{CI}) which gives about 70 per cent of the estimated valence shell correlation energy. Values close to the minima and maxima are shown for the SCF with a BSSE correction (E^{B}_{SCF}) after the counterpoise method of Boys and Bernardi [**81**], and BSSE-corrected CI energies (E^{B}_{CI}) using a method suggested by Saunders. [**82**] Size-inconsistency corrections were also calculated for the E_{CI} and E^{B}_{CI} energies using the Davidson formula [**83**]; on the scale of figure 15, these lie within 0.1 kcal mol^{-1} of the respective uncorrected values and hence are not shown separately.

The figure shows that the essential shape of the donor section is reproduced in all six sets of calculations; we conclude therefore that (within the limitations of constant monomer geometry assumed in the calculations) figure 15 is a good representation of the donor section *shape*. The same is not true either of *absolute* energy values or of relative well depths or barrier heights. The absolute scaling of the dimer energy is perhaps not too serious

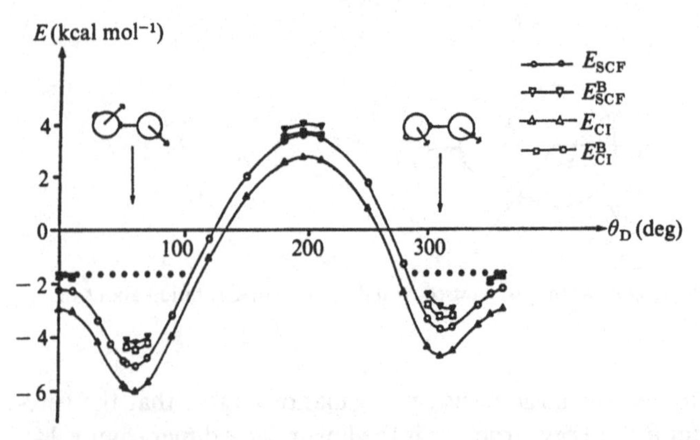

Figure 15. The donor section from quantum mechanical calculations at various levels (see text for basis set). The CI calculation includes all single and double excitations, except those from the lowest energy orbital of the monomer, the lowest two of the dimer, and those corresponding to the highest energy virtual orbitals. The dotted line is 3.5 kcal mol^{-1} above the E_{SCF} *trans* minimum.

a problem, as it can be fixed for a given model potential with respect to experimental data (e.g. second virial coefficient or potential energy of the liquid, or ice). The relative barrier heights and well depths are more problematical. Figure 15 shows us that the various corrections are not constant over the whole section, resulting in energy differences which are significant with respect to kT. Consider for example the *cis*→*trans* barrier in this energy surface section: depending upon the calculation and the correction applied, this can vary between 2.6 and 3.5kT, a difference which, in a room temperature ensemble, we would expect to lead to significant differences in relative populations of related configurations (see below section 5.3). Bearing in mind the uncertainty in these corrections, we are thus forced to conclude that an unavoidable uncertainty remains in the quantum mechanical results. In our use of the quantum-mechanical reference data to consider aspects other than surface section shape, therefore, we allow an uncertainty of up to about 2 kcal mol^{-1} in absolute energy values, and up to about 0.5 kcal mol^{-1} ($\sim kT$) in relative barrier heights and well depths. For the angular variables, we believe the uncorrected SCF reference data are adequate for our comparisons of shape and they are used in the immediately following discussion. For the distance variable r_{OO}, this is clearly inadequate, as electron correlation is essential to any consideration of the dispersion attraction. For the distance sections, therefore, the corrected CI reference data E^B_{CI} are used. These points are discussed more fully in references 32, 79 and 80.

5.3. *General Surface Features*

Figure 16 shows the ST2 and MCY $E(\theta_A, \theta_D)$ surfaces, and illustrates some of the features of interest. ST2 exhibits a double minimum, which reflects the

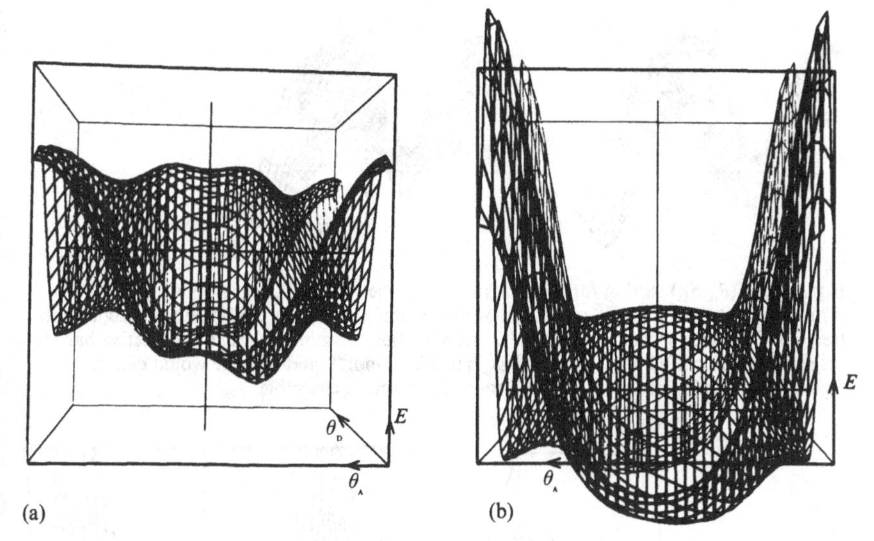

Figure 16. Three-dimensional dimer energy surface section $E(\theta_A, \theta_D)$ for (a) ST2, (b) MCY potentials. r_{OO} is fixed at 3.0 Å, $\psi = 0°$.

strong tetrahedrality of the model through a separation of the *cis* and *trans* minima, each of which corresponds to one of the two 'lone pair' charges pointing towards the donor proton. MCY is a three-point charge model, without such negative charge separation, so only a single broad minimum is observed, biased towards the *trans* form presumably largely because of proton–proton repulsion. The second feature to note is the height of the repulsive wings, which are quantitatively very different for the two models. The very high wings for MCY, with a maximum at about 20 kcal mol^{-1}, is probably due to a variety of connected reasons, possibly including inadequate flexibility of the functional form, and a relatively small number of relevant repulsive configurations with short H–H distances in the data base to which the form was fitted.[32] Of itself, such a large repulsion may have little effect on ensemble average structure, as it will not be sampled at room temperature.

Close inspection of figure 16 suggests possible low-energy pathways through the surface, though they are difficult to see in the figure. Figure 17 shows a stereo view of the same plot for SPC from a different viewpoint; a low energy path, corresponding to simultaneous rotations of θ_A and θ_D as shown in the motif of figure 17, exists on this surface. The barrier along this path is sufficiently low, in all potentials for which it has been examined, to be sampled in a room temperature ensemble. Such a pathway (which passes through a 'bifurcated hydrogen bond' at the top of the barrier) is likely to be important in a simulation and relevant to the dynamics of the assembly. The relative height of this barrier for different potentials will affect how different hydrogen bonding configurations can interconvert. It has been little investigated so far, and clearly deserves more attention.

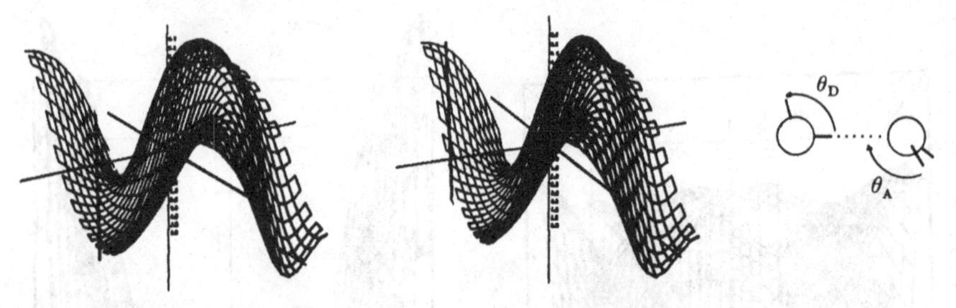

Figure 17. $E(\theta_A, \theta_D)$ section for SPC looking down the 'channel', corresponding to simultaneous rotations of θ_A and θ_D from the minimum dimer to a bifurcated configuration (see motif). A similar channel is found in all potentials examined, though the relative barrier heights vary. For SPC, the barrier is less than 3 kcal mol^{-1}, and thus we would expect significant sampling of this region in a room temperature ensemble.

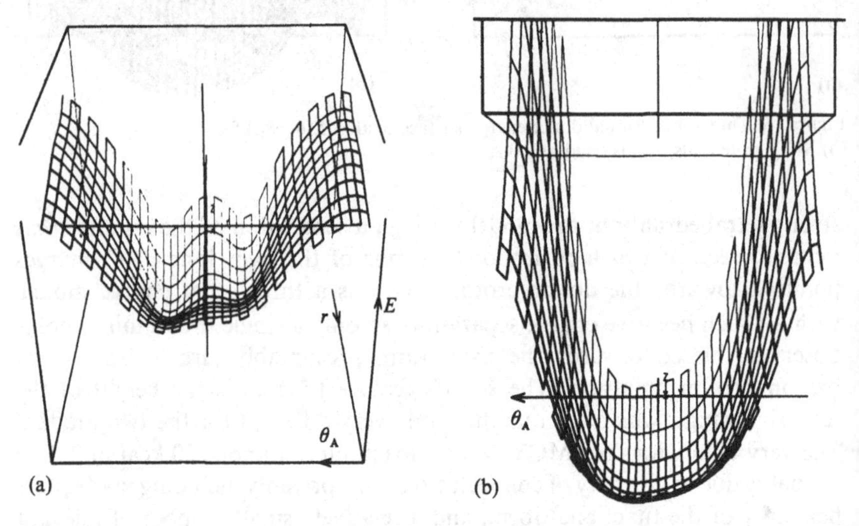

Figure 18. $E(r_{OO}, \theta_A)$ for (a) ST2, (b) MCY.

The behaviour of these minima and low energy barriers as a function of O–O separation are clearly of interest; one can perhaps already begin to suggest that the liquid structure may be strongly influenced by a balance between the reduction of such rotational barriers by small increases in O–O separation, compensated by the energy required to perform that increased separation. Thus perhaps we should pay particular attention to the $E(r_{OO})$ behaviour. Figure 18 shows $E(r_{OO}, \theta_A)$ for ST2 and MCY, and we can see how, as r_{OO} increases, the ST2 double well begins to transform into an asymmetric single well, with the intervening barrier passing through a point of inflection. The MCY curve retains its single-minimum feature, even at low r_{OO}; considering the three point-charge model underlying the functional form, this is not surprising.

5.4. *Principal Energy Surface Sections*

It is unrealistic to think of computing quantum mechanical surfaces to compare against the model curves shown in figures 16–18; computational restrictions force us to limit the region of the potential energy surface over which comparisons are made. Specifically, we concentrate on three sections of the six-dimensional surface, namely (a) $E(r_{OO})$, the distance section, in which θ_A and θ_D are kept constant at 'standard values' at (or near) their minimum dimer values; (b) the acceptor section $E(\theta_A)$, in which the acceptor molecule only is rotated (figure 14), (c) the donor section $E(\theta_D)$ in which the donor molecule is rotated.

5.4.1. *Problems with Comparing Energy Surface Sections*

(*a*) *Choice of configuration.* Different potentials give global minima at different values of R_{OO}, θ_A, and θ_D. In comparing two or more model potentials with one another, we should ideally compare sections through the global minimum which give information on, for example, possible low energy paths (e.g. the 'channel' discussed above), the steepness of the well walls, and any possible rotational barriers. Even for similar potentials, 'equivalent' sections may lie along slightly different directions in our six-dimensional space. Although it is in principle possible to select such equivalent sections, it is a complex procedure, and requires probing all six variables. In comparing a model potential with the quantum mechanical reference data, it is not computationally feasible. We have chosen, therefore, a compromise procedure.

Our comparisons with quantum mechanical reference data use a 'standard configuration' which is that of the linear dimer $\theta_D = 52.26°$, $\theta_A = 120°$. In this orientation, the donor proton points directly towards the acceptor oxygen. For an ideal tetrahedral orientation of the acceptor, we would choose an acceptor angle of 125.17°. The limited significance of this ideal direction implies only a weak argument for using it; moreover the general flatness of the acceptor angle section (e.g. see figure 21(a)) suggests the choice of a standard θ_A is not critical. For the model potentials used, θ_{Amin} varies from 125° to 160°, all of which values fall within the quantum mechanical well of figure 21(a). Within a given *group* of potentials, the variation of θ_{Amin} is much smaller, and never more than $\pm 5°$; the effect on the energy of these θ_A differences is negligible. Similarly, θ_{Dmin} values generally fall within 10° of 52°, the only exception being PE($\mu\Omega$) ($\theta_{Dmin} \sim 69°$). Despite the sharper $E(\theta_D)$ well (figure 19), the effect of this displacement on the dimer energy is also small in all cases.

In making comparisons between potentials, sections are chosen for which the fixed angular variables are at their minimum values where possible. However, use of sections at slightly different fixed θ_A and θ_D values for the donor and acceptor sections respectively, leads to the same results, confirming

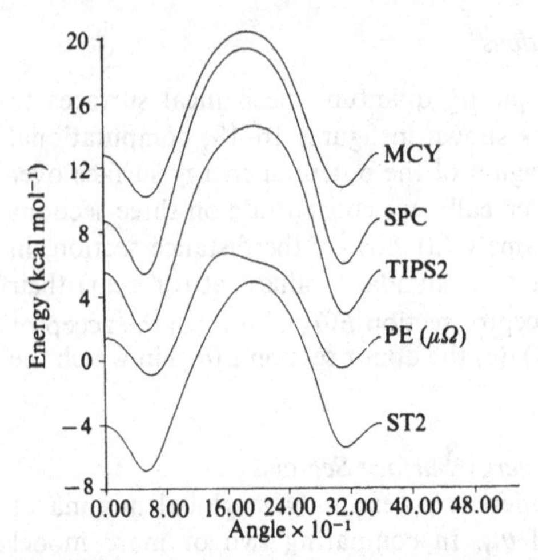

Figure 19. Donor sections $E(\theta_D)$ of the dimer energy surface for the models indicated. The similarity in shape of all models should be noted. Configuration: $\theta_A = \theta_{Amin}$, $r_{OO} = r_{OOmin}$. Energy zero of each successive section is shifted by 4 kcal mol^{-1}.

that small differences in the choice of the fixed angular variable(s) do not affect the conclusions drawn. The choice of r_{OO} is much more critical; we invariably take it at the minimum dimer value for the standard configuration.

(*b*) *Significant energy levels.* In liquid water we are concerned with an ensemble average at room temperature, for which the typical thermal fluctuation kT is of the order of 0.5 kcal mol^{-1}. Thus, we are interested not specifically in the absolute minima in these sections (which in any case are not, with one exception, minima in the full six-dimensional surface) but in how the total surface is sampled by an assembly at that temperature. Thus, energy gradients, low (compared to kT) energy barriers, and shapes of energy wells are important, and differences in these features of the order of kT between different models may lead to significantly different ensemble average structures, which themselves exhibit the results of a complex interplay of orientation and distance dependencies. We are not in general interested in energy barriers many times kT in height, as such regions would not be sampled in a room temperature simulation. Thus the anomalously high repulsive wings of MCY may be of little interest in themselves, although if one consequence of such high wings is a narrowing of the acceptor well, then this will restrict the range of θ_A values that are sampled.

What range of energies is, therefore, significant, and which regions of the potential energy surface will be sampled most heavily? We can get some indication of this range from simulation studies. Mezei and Beveridge [**84**] plot nearest-neighbour pair energy distributions for several potentials, the full width at half height of the distributions showing energies up to 3–4 kcal mol^{-1} above the minimum are well-sampled. A rough curve fitting to the *total* pair energy distributions for ST2-water[50] show a similar half-width of about 3 kcal mol^{-1}, with an upper energy value of about 3.5 kcal mol^{-1}. Thus, the region between the global minimum and an energy about 3.5 kcal mol^{-1} higher is particularly important for room temperature simulations. An energy level about 2 kcal mol^{-1} above the minimum is at about the centre of the pair energy distributions; dimer configurations of this energy might thus be expected to be of particular importance.

An additional uncertainty does arise from quantum corrections. Although most of our models are of the rigid body type, there is still the zero point motion of the O···O intermolecular stretching vibration to consider. From quantum mechanical calculations of the $E(r_{OO})$ section, this frequency is estimated (and measured) to be about 150 cm^{-1} giving $h\nu/kT$ approximately equal to 0.38 kcal mol^{-1}. In order for a vibration to be treated classically, $h\nu/kT$ should be much less than unity; this is not so in this case, so a zero point energy term of approximately 0.2 kcal mol^{-1} may be relevant. Compared to the energy levels discussed in the preceding paragraph, this can be neglected.

In the following discussion, therefore, we will be interested primarily in the structure of the energy surfaces up to about 3.5 kcal mol^{-1} above the minimum, paying particular attention to the 2 kcal mol^{-1} level.

5.5. *Donor Section*, E(θ_D)

Figure 15 shows the SCF reference section; from figure 14 we see that the two minima in the section correspond to *trans* and *cis* configurations, with the *trans* being the deeper minimum by about 1.4 kcal mol^{-1}. The shallower *cis* minimum can be rationalized in terms of the greater proton–proton repulsions in this configuration. The minima are separated in this section by two maxima. The lower barrier ($\sim$ 2.8 kcal mol^{-1} above the deeper minimum) corresponds to the donor molecule's dipole pointing directly along the OO line, and would be encountered when, for example, a hydrogen-bonding arrangement involving one hydrogen of the donor switches to the other by rotation. The second can be rationalized in terms of lone pair–lone pair repulsion. The +3.5 kcal mol^{-1} level is marked, and suggests that the shape of the surface in the region of the lower rotational barrier may be critical for the ensemble average structure. Even for a hypothetical dimer gas, differences in the relative depths of the two wells, and the intervening barrier, would lead to different average configurations at room temperature. Thought of as a simple two-state model, a change in the difference between the two well depths from, say 1 to 1.5 kcal mol^{-1} would have a significant effect on the relative populations of the two states. Thus it seems important that an adequate potential function should be able to reproduce the 'correct' detail of this region to much better than kT.

As this is merely a section of a six-dimensional surface, the sampling in this region will depend also upon the shape of the neighbouring region in the other dimensions; on this, we have little information. In this context, we should note that the *cis* minimum is only an apparent minimum for most potentials because of the restricted dimensionality of the plot; there is a direct path to the global minimum in another dimension – namely a change of θ_A (figure 21) to give the *trans* configuration.

Figure 19 shows the donor sections for five representatives of the major types of models, *viz.* four-point-charge five-centre (4/5, e.g. ST2), three-point-charge three-centre (3/3, e.g. SPC), three-point-charge four-centre (3/4, e.g. TIPS2), Clementi form (MCY, e.g. MCY), and PE (e.g. PE($\mu\Omega$)). In these sections, θ_A and R_{OO} have been set to their respective minima. We note immediately the strong similarity in form both between the sections, and with the quantum mechanical reference section (figure 15). This similarity is very striking, especially the near-quantitative similarity for lone pair–lone pair repulsive configurations (the central barrier), even though their modes of treatment of the lone-pair regions are not uniform.

Figure 20 attempts to assess the quantitative differences between these

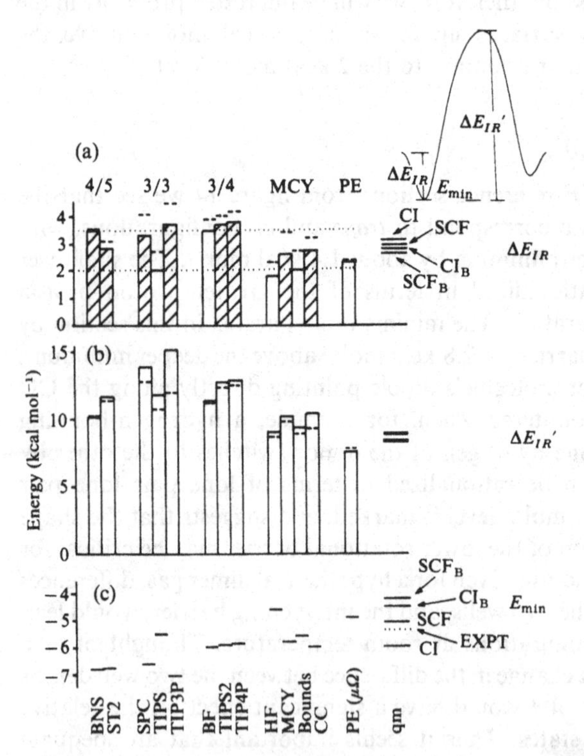

Figure 20. Relative apparent minor (a) and major (b) barrier heights ΔE_{IR} and $\Delta E_{IR'}$, and (c) absolute well depths for donor sections (figure 19). QM data and broken lines relate to $\theta_A = 120°$, full lines to $\theta_A = \theta_{A\min}$. Dotted lines in (a) show the 2 and 3.5 kcal mol^{-1} levels. Shading in (a) indicates potentials which give water-like g_{OO}s (table 7).

donor sections, grouped according to type. We pay particular attention to the heights of the two *trans*→*cis* barriers ΔE_{IR} and $\Delta E_{IR'}$. Also shown are the absolute minima of the potentials.

Considering first the absolute minima, there is a broad distinction between the EPPs and true dimer potentials in that the former tend to be significantly deeper than both the experimental and BSSE-corrected quantum mechanical values; the true dimer potential values tend to oscillate closer to the experimental. This separation of behaviour is not complete, for example the TIPS minimum is anomalously shallow compared to the other 3/3 models; we can, however, reasonably generalize that the deeper EPP minima probably reflects an allowance for non-additivity. The anomalously shallow HF well reflects the absence of an attractive dispersion term.

Turning to the major barrier $\Delta E_{IR'}$, although there are differences between the potentials, all lie well above the 3.5 kcal mol^{-1} level, and so differences here are unlikely to be significant at normal temperatures and pressures. PE($\mu\Omega$) is significantly lower than the others. Because the quantum mechanical section refers to $\theta_A = 120°$, comparisons with this reference data are problematical. The broken lines give the model potential barriers at this value of acceptor angle; even bearing in mind our qualifications concerning the *quantitative* reliability of the quantum mechanical results, the 4/5, 3/3 and 3/4 types all show barriers that are too high. Whether this disagreement would be reduced by suitable rotations of potential function surfaces to make them more nearly 'congruent' with respect to their major topographical features is not clear.

The minor barrier heights show two types of behaviour. Apart from TIPS, which has an anomalously low barrier, the 4/5, 3/3 and 3/4 types all have heights between about 4 and 5 kcal mol^{-1}. In contrast, the MCY and PE types are significantly lower at about 3–4 kcal mol^{-1}. Again, comparison with the quantum mechanical data is problematical; considering the (broken line) model potential values for the standard configuration ($\theta_A = 120°$), there is a suggestion that the 3/3 and 3/4 types have barriers that are probably a little too high.

What general statements can be made regarding these donor section differences? First, for the EPPs there seems to be a tendency not only to lower minima, but also higher barriers than we might expect from the reference data. The minor barriers being at heights comparable to our 2 kcal mol^{-1} level implies differences in ΔE_{IR} between different potentials will be significant in determining relative populations of different configurations in the neighbourhood of the minimum dimer configuration. More specifically, we would expect a higher barrier for those potentials *within a group having similar form* to be reflected in stronger orientational correlations in the liquid. We may anticipate section 6 at this point in noting that those potentials which give water-like behaviour are in fact those with the highest barriers within a group, and are shaded in figure 20. As different functional forms will have potentially different (albeit apparently only slightly) inherent orientational energy dependencies, we cannot make similar generalizations *between* the different groups.

5.6. *Acceptor Section*, $E(\theta_A)$

Figure 21 (a) shows the SCF acceptor section. There is only one minimum, corresponding to the *trans* dimer; the *cis* dimer – which has a definite minimum in the donor section – is on the rising part of the curve, and hence is unstable as a dimer configuration.

A selection of acceptor sections for various models is given in figure 21 (b). These are given for r_{OOmin} and $\theta_D = 52.26°$. We immediately see shape differences that are characteristic of the different model types. ST2 shows clear double-well features, reflecting the separation of negative charge in four-point charge models; the quantitative reduction in tetrahedrality in going from BNS (not shown) to ST2 is reflected in the reduction of this double feature. Similar double-well behaviour is found in Jorgensen's four-point charge model. **[52]** PE($\mu\Omega$), and EPEN/2 (not shown) both retain remnants of this behaviour, though with no clear separation of two minima. The relative strengths of these two features can be rationalized in terms of the relative magnitudes of the dipole and quadrupole moments, as shown in figure 22. The *cis* configuration is favoured by the dipole–quadrupole interaction, while the *trans* minimum is dominated by the dipole–dipole term, with some dipole–quadrupole contribution. The relative depths of the two wells would thus vary with the relative magnitudes of these two leading terms, which as mentioned in section 3.6, will change when non-pair-additivity is taken into account in polarizable models.

The three-point charge models all show single minima, but with interesting differences. TIPS2 (a 3/4 model with the negative charge and Lennard–Jones centres separated) has a flattish minimum, while SPC (where the negative charge and LJ centres coincide) is much more convex. One consequence of this is the shift of the θ_A minimum *well* away from the tetrahedral to around 160°: the fact that a water-like g_{OO} can be obtained with this potential (see section 6.1) underlines the earlier discussion concerning the relative unimportance of tetrahedrality in the potential function itself. The MCY well has a shape similar to TIPS2, though narrower and with *much* higher wings relating to close proton–proton approach. The central force model (not shown) has *two* very high repulsive regions ($\sim$ 50 kcal mol^{-1} above the minimum) with a secondary minimum between them. As in MCY and its relatives, this region would be insignificantly sampled in a room temperature simulation, so it should not be regarded as a serious drawback in practice. The bottom of the CF well is also approximately linear, but over a much greater angular range ($\sim$ 140°) than any of the other models.

Table 5 gives the relative magnitudes of the acceptor-rotation barrier height (corresponding to the acceptor dipole pointing directly at the acceptor proton,

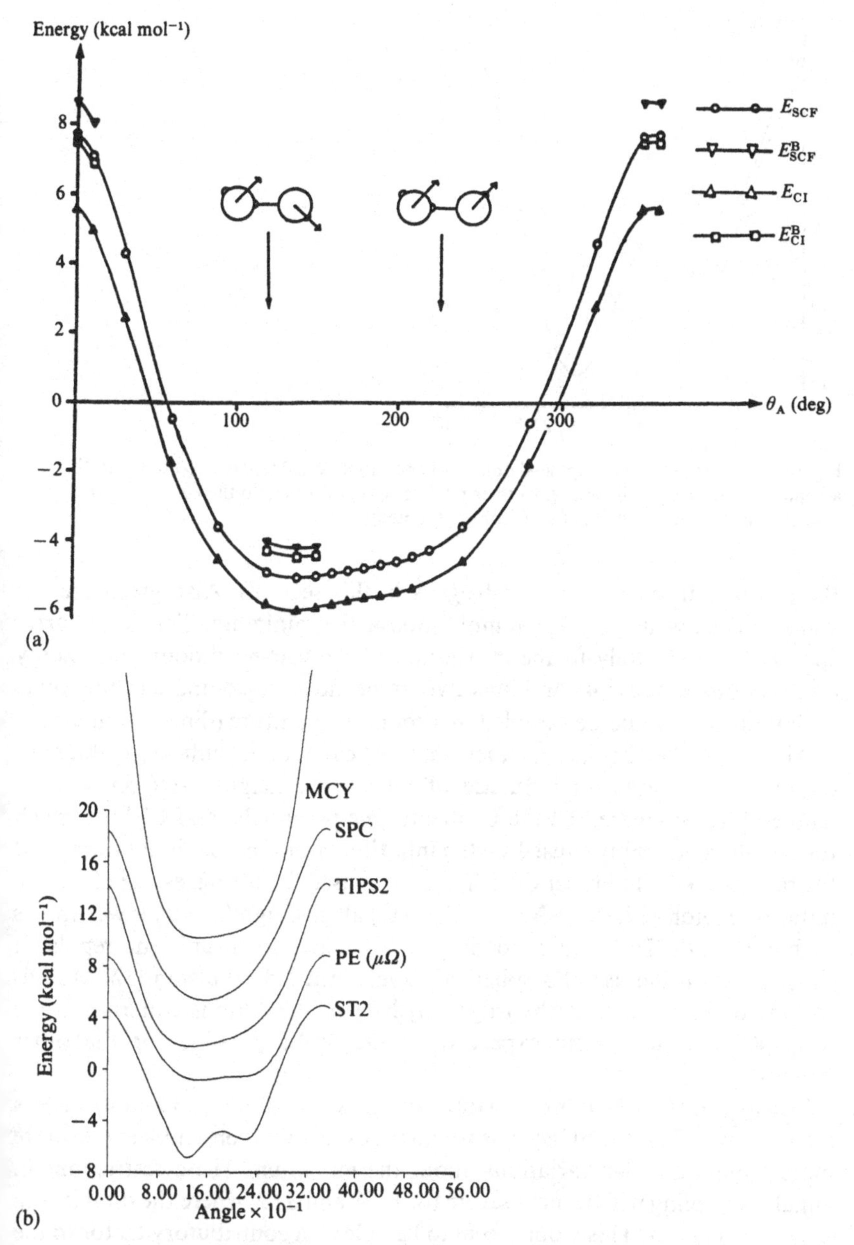

Figure 21. Acceptor sections $E(\theta_A)$ for (a) quantum mechanical calculations specified in the caption of figure 15, and (b) five specified models. Note the overall similarity in the well shapes, but detailed differences for the different types of potential (see text).

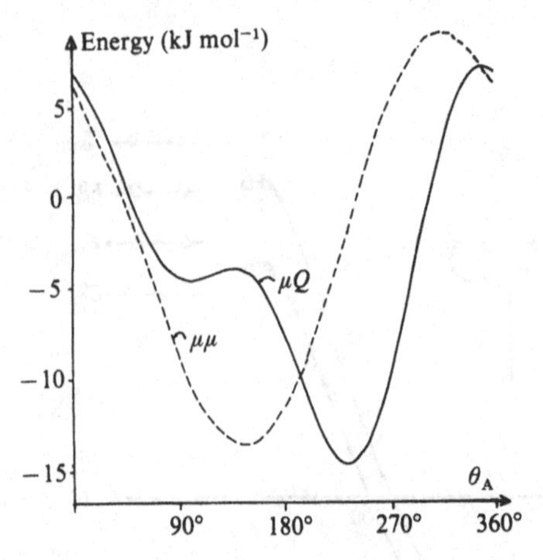

Figure 22. Contributions of dipole–dipole ($\mu\mu$) and dipole–quadrupole (μQ) terms to the acceptor section $E(\theta_A)$, showing how their relative strengths relate to the *cis* and *trans* configurations. Data plotted is for the PE(QQ) potential.

the closest approach of the hydrogens in this section). Also given are the widths of the wells at 2.0 kcal mol^{-1} above the minimum. This level corresponds approximately to the maximum in the near-neighbour pair energy distribution (section 5.4), and thus gives some indication of the acceptor angle variations that would be sampled in a room temperature (dimer) simulation.

All barrier heights are greater than 9 kcal mol^{-1}, sufficiently high to discount any significant influence of the barrier height itself on a room temperature ensemble. All the Clementi-type potentials (and CF) are much too repulsive, possibly caused by the inflexibility of the functional form, and the relatively small number of sample points used for fitting, especially in the repulsive region. [**32**] BF, BNS and PE ($\mu\Omega$) all give significantly lower values than the SCF. This could possibly be assigned to insufficient repulsion consequent on the use of a spherical oxygen-centred repulsion. [**32**] Modifications to PE to include hydrogen–hydrogen interactions can raise these wings [**43**] and we would expect the same to be possible for the other potentials.

Turning to the 2 kcal mol^{-1} widths of the wells, all the 3/3 and 4/5 EPPs are narrower than the SCF; thus we might expect their ensembles to explore more limited angular variations about the minimum. These restrictions in angular sampling will be most severe for BNS and ST2, where the double well structure remains. This would seem to be at least a contributory factor to the strong tetrahedrality of BNS and ST2-water. The Stillinger–David polarization model has an even narrower θ_A well; like SPC and TIPS, it has no linear region near the well bottom.

The Clementi-type potentials almost all have about the same width as the SCF, MCY being slightly narrower, CC slightly wider. PE($\mu\Omega$) and Rowlinson

Table 5. *Acceptor section characteristics*[a]

	Height of barrier above minimum (kcal mol^{-1})	Width of well at 2 kcal mol^{-1} above minimum	Type of potential
SCF[b]	12.9	165°	
SCF+BSSE[b]	12.9	*	QM
CI[b]	11.7	165°	
CI+BSSE[b]	12.2	*	
BNS	10.3	140° overall (50°+30°)[c]	4/5
ST2	11.8	140°	
SPC	13.2	140°	
TIPS	12.5	140°	3/3
TIP3P	14.2	135°	
BF	10.6	185°	
TIPS2	12.5	170°	3/4
TIP4P	12.3	170°	
RWK2	*	~ 190°[d]	
HF	19.3	165°	
MCY	26.9	160°	
Bounds	21.8	165°	MCY
CC	30.4	170°	
PE ($\mu\Omega$)	9[6	190°	
SD[e]	*	~ 100°[d]	Polarizable
CF	47.2	150°	Non-rigid body
SS[f]	*	> 240°	EPEN/2

* Not calculated.

[a] For minimum distance (defined in standard configuration) and $\theta_D = 52.26°$

[b] For $r_{OO} = 3.0$ Å. Minimum distance varies between 2.95 and 3.05 Å. Gradient near minimum implies an error in barrier height will be less than 0.1 kcal mol^{-1}.

[c] At the 2 kcal mol^{-1} level, the *trans* and *cis* wells are separated; their widths are 50° and 30° respectively.

[d] Width estimated from plots in reference 61.

[e] Stillinger–David polarization model.

[f] Shipman-Scheraga EPEN/2 parameters.

(not shown) acceptor wells are 25–30° wider, suggesting that simulations using these potentials will tend to be insufficiently orientationally restricted. EPEN/2 (SS) is even wider, by at least an additional 50° over PE and Rowlinson. All the 3/4 model wells are wider than the quantum mechanical ones, though it is marginal (~ 5°) for TIPS2 and TIP4P. BF and RWK2 are 20–25° wider.

These differences do begin to make some kind of sense in the context of structural results from simulation calculations: as we shall see, water-like behaviour correlates to some degree with narrow θ_A wells. However, the discussion so far has considered orientational energy dependencies for a fixed separation distance only. Before considering structural results in more detail, we examine the $E(r_{OO})$ section.

5.7. *Distance Section*, $E(r_{OO})$

5.7.1. *Comparisons with Quantum Mechanical Reference Data*. In comparing the angle-dependent energy-surface sections we were concentrating on orientational dependence of the dimer energy at the equilibrium dimer separation and could (almost) justifiably ignore the problem of non-pair-additivity. When we come to considering distance-dependence, however, this is no longer possible. We need to begin to make a specific distinction between effective pair and 'true dimer' potentials.

We would expect the $E(r_{OO})$ section of an EPP to differ from the reference dimer section in two respects. First, the expected overall attractive consequence of non-pair-additivity will lead to a shorter first-neighbour distance than for the dimer (as found in liquid water and ice), and hence the overall well will be shifted to shorter distances. Secondly, the additional binding energy will lead to a deeper well than for the case of the isolated dimer (as seen in figure 20). Thus, we should not expect the $E(r_{OO})$ sections of EPPs to be directly comparable with the quantum mechanical reference data.

A second problem concerns the quantum mechanical reference data to be compared with model potentials. As discussed earlier, SCF data will neglect the dispersive attraction term, and hence be inadequate. As shown in figure 23(a), which gives the quantum mechanical results for the 'distance well', the progression from (uncorrected) SCF to CI significantly deepens the well, as would be expected from the addition of a further attractive component (with leading term $\propto r^{-6}$). Secondly, the actual position of the minimum is strongly dependent upon both the inclusion of dispersion through the CI term, and the BSSE correction: a shift of up to 0.1 Å can thereby be induced, which is outside the error of the experimental dimer separation.[**79**, **80**, **85**] Thirdly the inclusion of BSSE in both SCF and CI cases reduces the well depth. The uncertainty concerning the BSSE correction thus prevents us from relying on the quantum mechanical results for *quantitative* information on either the well depth or the details of its shape. This is particularly unfortunate, as (see later discussion) the shape and width of the $E(r_{OO})$ well appears to be crucial to the success or otherwise of a model potential function. In this section, we use the BSSE-corrected CI curve as reference data, though we should bear in mind the relatively high uncertainty in the quantitative validity of this.

5.7.2. *Model Potential Function Distance Wells*. All model potentials yield $E(r_{OO})$ wells that look similar in form to figure 23(a); some examples are given in figure 23(b). There are, however, considerable quantitative differences

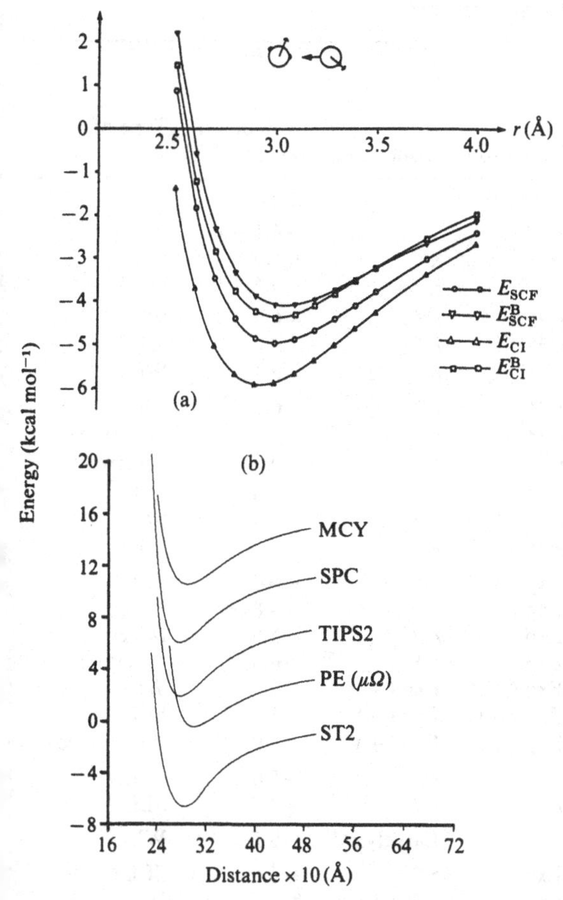

Figure 23. Distance sections $E(r_{OO})$ for (a) quantum mechanical calculations specified in the text (E^{B}_{CI}), and (b) five potential functions. Standard configuration: $\theta_A = 120°$, $\theta_D = 52.26°$.

between them. Table 6 summarizes the repulsive and dispersion forms, and gives an estimate of the *width* of the well at 2 kcal mol^{-1} above the minimum energy.

Turning first to the EPPs, all models give distance wells that are *deeper* and *narrower*, and at lower r_{OO} than the reference data imply. The depth and r_{OO} changes are expected from our considerations of the 'overall' effect of non-pair-additivity; the *narrowness* of the wells is not necessarily expected from a consideration of the likely effects of non-pair-additivity. In this context we should note that the *broadest* well is that of TIPS, and we shall see in the following section that this potential, of all the EPPs, is alone incapable of yielding reliably a second g_{OO} peak at about $1.6r_0$. We should also stress that all these wells are clearly narrower than implied by the quantum mechanical

Table 6. *Distance section characteristics*[a]

	Width at 2 kcal mol^{-1} above minimum (Å)	Repulsion	Dispersion	Well depth (kcal mol^{-1})	Type of g_{OO}[b]
SCF	1.09	—	×	−5.0	—
SCF+BSSE	1.32	—	×	−4.1	—
CI	0.98	—	√	−5.9	—
CI+BSSE	1.12	—	√	−4.4	—
BNS	0.80	r^{-12}	r^{-6}	−6.9	WL
ST2	0.70	r^{-12}	r^{-6}	−6.8	WL
STO-3G	0.57	r^{-12}	r^{-6} (+)	−6.4	WL
SPC	0.78	r^{-12}	r^{-6}	−6.6	WL
TIPS	0.96	r^{-12}	r^{-6}	−5.7	SLL (mixed?)
TIP3P	0.75	r^{-12}	r^{-6}	−6.5	WL (weak)
BF	0.76	r^{-12}	r^{-6}	−6.0	SLL
TIPS2	0.80	r^{-12}	r^{-6}	−6.2	WL
TIP4P	0.76	r^{-12}	r^{-6}	−6.2	WL
HF	1.28	Exp	×	−4.4	SLL
HFW	1.16	Exp	r^{-6}	−4.8	*
HFL	1.02	Exp	r^{-6}	−5.6	Mixed
HFK	0.92	Exp	r^{-6}	−6.5	Mixed
MCY	1.02	Exp (+)	Exp (+)	−5.5	WL
Bounds	1.09	Exp (+)	Exp (+)	−5.3	SLL
CC	0.90	Exp (+)	Exp (+)	−5.9	WL (mixed)
PE(QQ)	0.93	r^{-12}	r^{-6}	−5.0	SLL
PE($\mu\Omega$)	1.03	r^{-12}	r^{-6}	−4.5	SLL
CF	0.54	r^{-n}	Largely r^{-6}	−6.5	WL
SS	0.99[c]	Exp	r^{-6}	−5.5	SLL
QPEN	*	Exp	r^{-6}(+)	*	WL

[a] In standard configuration ($\theta_A = 120°$, $\theta_D = 52.26°$), except where stated.
[b] WL: water-like; SLL: simple-liquid-like.
[c] At configuration in original paper.
* Not calculated.
(+) and anisotropic.

reference data, even when the likely errors in the latter are allowed for. The CF well is anomalously narrow, being approximately 50 per cent of that of the reference data.

Considering now the Clementi-type 'true dimer' potentials, we recall first the strong similarity between their $E(\theta_A, \theta_D)$ sections discussed above. The pure HF potential – which is completely devoid of a dispersion attraction – is *very* broad; as various dispersion terms are added, the well breadth reduces as expected. Intermediate in terms of width are MCY and the Bounds reparametrization; these give well-breadths of the order of that indicated by the quantum mechanical reference data. We note in particular that the effect of the Bounds reparametrization is to *increase* the width by some 7 per cent. The EPEN and PE(QQ) widths are perhaps slightly too narrow; adding the dipole–octupole term to give PE($\mu\Omega$) increases the width by some 10 per cent. The STO–3G potential of Jorgensen gives an *extremely* narrow well indeed. The CC and CH potentials (not shown) have significantly different distance well shapes, the CH potential having an additional $re^{-\alpha r}$ dispersion term in the fitted form. In general, the EPPs fall off more rapidly with distance than the 'true dimer' potentials.

We postpone further discussion to the following sections, in which we find that the narrowness of the well correlates strongly with the water-like nature of the simulated structure.

6. Simulated Liquid Structures

As discussed in section 2, liquid water has an oxygen–oxygen pair correlation that is distinctive, and qualitatively different from the inert gas liquids. We refer again to figure 1 and note the relative positions of the second peak at approximately $1.6r_0$, and $2r_0$ for 'water-like' (WL) and 'simple liquid-like' (SLL) PCFs respectively. Clearly, an adequate water potential should have a water-like g_{OO} as a first test. A second essential requirement stressed in sections 2 and 3 is quantitative agreement with experiment of the peak positions, peak heights, depths of minima, and peak areas (which relate to coordination number). As mentioned in section 2, there remain some uncertainties in the experimental data, and still insufficient understanding of the effects of ignoring the non-rigid-body nature of the water molecule, and its associated quantum effects. We might expect these effects to be less for g_{OO} than for g_{OH}. Partly for this reason, partly because of greater experimental uncertainties in the other partials (see **[9]**), and partly because we expect g_{OO} to give clearer information on orientational structure (section 2), we limit our discussion largely to g_{OO}.

6.1. *Effective Pair Potentials* (*EPPs*)

We refer again to table 6, concentrating initially on the nature of the g_{OO} obtained from liquid phase computer simulations. We note immediately that all the EPPs except TIPS clearly give water-like g_{OO}s, though the relative heights of the various peaks are very variable (table 7). We note also that all the repulsions are r^{-12} and hard, while for all those for which estimates have been made, the 2 kcal mol^{-1} distance-well width is narrow, in fact some 70–80 per cent of that expected from the CI reference data. The EPP with the widest well (TIPS, approximately 80% of the corrected quantum mechanical CI width) is interestingly the only one that fails to give a water-like g_{OO} at all, though the second peak for TIP3P is weak. **[72]** We thus have some suggestion that a narrow distance well is conducive to a water-like g_{OO}.

Although the *form* of the g_{OO} is correct for most of these EPPs, a closer look at the PCFs reveals significant disagreement with experiment. Table 7 lists the heights and positions of the first two g_{OO} peaks (h_1 and h_2) and the first minimum (m_1), together with the estimated first coordination number, and the h_1/h_2 ratio. We can make the following generalizations.

(a) Although the forms of the computed g_{OO}s are correct, *the relative heights of the first two peaks are in significant disagreement with experiment.* Taking the Narten experimental data as reference, the heights of the computed first peaks range from 13 per cent (BNS) to 65 per cent (CF) too large, way outside the expected experimental errors. We note with interest that the progression from BNS to ST2 *worsens* the agreement of the first peak height with experiment from 13 per cent to 32 per cent too high, *though the height of the second peak* (h_2) *improves.* For most EPPs the heights of the second peak are too low.

(b) *There seems to be some correlation between the heights of the first and second peaks* as indicated by the h_1/h_2 ratios of many of the models. Apart from BNS, h_1/h_2 is approximately the same for all the EPPs listed except for BNS and CF, and this ratio is some 30 per cent larger than that derived from Narten's experimental data (and even larger compared to the Dore data.[**86**]) Considering both four-point charge models, and three- or four-centre three-point-charge models are represented, this approximate constancy is interesting. This h_1/h_2 ratio underlines a further common feature that has been found with these water potentials: in order to obtain a $1.6r_0$ peak, we seem to be forced to modify the potential so as to enhance the first peak to well above the experimental value. Both these points tempt a speculation that there may be something basically wrong with these model representations of real water which all have similar orientational dependencies. We might note also a rough correlation between the height of the first peak and the width of the distance well discussed in the previous section: the narrowest well (CF) gives a very high (65 per cent too high) first peak (and incidentally a reasonable second peak height!). As we increase the distance well width, h_1 falls, but so unfortunately does h_2. By the time the well has been widened sufficiently to give a reasonable h_1 (see the TIPS entry in table 7), we lose the second peak completely (and increase the area under the first peak unacceptably). Thus, for potentials of these EPP forms, forcing a quantitatively-correct second peak at $1.6r_0$ by narrowing the distance well enhances h_1 too much. Again, this suggests there are significant problems with the basic potential which need attention. Wider implications are discussed later: it seems the problem is unlikely to be solved with potentials showing this kind of orientational dependence, whose influence on the directional structure is enhanced by narrowing the width of the distance well.

(c) The four-point charge, and CF models give too low a first minimum. This implies a first neighbour shell that is too orientationally structured. All models apart from ST2 underestimate the depth of the *second* minimum (not shown), and hence seem to account inadequately for longer range correlations.

Table 7. *Characteristics of water-like* g_{OO}*s*

Potential	Distance[a] well width (Å)	h_1 (r_1)[b]	h_2 (r_2)[b]	m_1 (r)[b]	h_1/h_2	r_2/r_1	CN_1[c]	Ref.[d]
BNS	0.80	2.68 (2.75)	1.23 (4.6)	0.54 (3.5)	2.18	1.67	5.3	61
ST2	0.70	3.13 (2.84)	1.18 (4.7)	0.66 (3.4)	2.65	1.65	5.1 (4.9)	61 (72)
SPC	0.78	2.86 (2.77)	1.08 (4.6)	0.84 (3.5)	2.65	1.66	5.1†	72
TIP3P	0.75	2.75 (2.81)	1.02 (4.6)	0.86 (3.6)	2.70	1.64	5.2†	72
TIPS2	0.80	3.04 (2.80)	1.15 (4.5)	0.79 (3.4)	2.64	1.61	4.8†	72
TIP4P	0.76	3.00 (2.76)	1.12 (4.5)	0.81 (3.4)	2.68	1.63	5.1†	72
RWK1	*	4.00 (2.79)	1.10 (4.5)	0.79 (3.3)	3.64	1.61	4.5	61
MCY	1.02	2.46 (2.83)	1.08[e] (4.25)	0.94 (3.5)	2.28[e]	1.50	4.7	61, 93[e]
CF	0.54	3.94 (2.85)	1.18 (4.4)	0.38 (3.2)	3.33	1.54	4.5	75
CF(mod)	*	3.05 (2.85)	1.11 (4.5)	0.61 (3.3)	2.75	1.58	4.5	75
Expt[f] (25°C)	—	2.38 (2.85)	1.14 (4.65)	0.80 (3.5)	2.09	1.63	5.3	12

* Not calculated.
[a] Width 2 kcal mol^{-1} above minimum (table 6).
[b] Heights of successive g_{OO} maxima (h_i) and minima (m_i), together with the relevant r_i value in brackets.
[c] Coordination number taken up to first minimum or (†) 3.5 Å.
[d] Source of data quoted. Reference 72 refers to (N, P, T) ensemble calculations at 25 °C and 1 atm, reference 61 refers to (N, V, T)
[e] All data except coordination number from reference 93. Reference 61 gives $h_2 = 0.96$, which is curiously low; the corresponding h_1/h_2 ratio is 2.57.
[f] Termination ripples make precise h_i and r_i location subject to error which is difficult to quantify; we estimate $\leqq 1$ per cent for r_1, r_2 and h_2.

This point has received virtually no attention; we suspect it might be important.

(d) Coordination numbers are problematical to compare, as their definition varies between workers. Four-point charge models (BNS and ST2) seem to give a coordination number about 25 per cent too high when measured out to the first minimum, but seem reasonable when taken out to 3.5 Å. This is presumably a consequence of too low a first minimum which relates to neighbour molecules which should be in this region being pulled in too close to the origin molecule by the enhanced distance attraction from the deep, narrow distance well. Partly because of these difficulties, we make no further coordination number comparisons, though we would underline it is essential to obtain a correct coordination number, measured in a consistent way: the areas of the various PCF peaks must be correctly reproduced by an adequate model.

(e) Except for ST2 and CF models, all first peak maxima occur at slightly too low an r value. This may seem a very minor point, as we might argue a simple distance renormalisation would correct this. A closer inspection of the r_2/r_1 ratios in table 7 suggests however that such a renormalization would in several cases move out the position of the second peak unacceptably. This is a particular problem for BNS and SPC, and again suggests that the relationship between the origin of the $1.6r_0$ peak and the position and height of the first peak needs further clarification and probably modification. These effects are perhaps on the borderline of being experimentally significant, but the relation between the relative peak *heights* h_1 and h_2 suggests they may repay looking into further.

6.2. '*True Dimer*' *Potentials*

Most of the 'true dimer' potentials produce simple liquid-like g_{OO}s. A notable exception is MCY, which gives a first peak of reasonable height, although the second peak is significantly low: the h_1/h_2 ratio is closer to experiment than any of the EPPs except BNS (table 7). A major problem is the positioning of the second peak at about 4.2 Å, corresponding to a *mean* acceptor angle of only 95°. This is a major disagreement with experiment, and implies the orientational dependence of the potential requires strengthening in some way.

Before discussing the true dimer potential results in more detail, we should notice an apparent discrepancy in table 6: *MCY yields a water-like PCF, yet the Bounds reparametrization, which is a better fit to the CI data points, gives simple liquid-like PCF.* These results are particularly interesting for several reasons. First, as the fit to the original CI data points is improved, the water-like nature of g_{OO} is reduced; not only does the second peak shift out to about 6 Å, leaving a minimum in the 4–5.7 Å region where the maximum was previously, but the position of the first maximum shifts out from approximately 2.8 Å (2.87 Å according to Bounds [**65**]) to approximately 2.96 Å. This is more in agreement with the dimer O–O separation predicted from the CI points themselves. Secondly, the first peak is broadened to give a trailing edge which falls off more slowly with distance; the height is about the same, implying that at least some of the neighbours previously at approximately 4.5 Å have shifted inwards. This is what we would expect to happen if the orientational dependence of the potential were weakened, either by a weakening of the orientational part of the potential itself, and/or a weakening of the O–O attraction, allowing the assembly to explore a greater angular freedom at slightly larger separations. The Bounds PCF (figure 24) in fact bears a strong resemblance to the PCF obtained from calculations on the MCY liquid under pressure; the suspected breakdown in the orientational correlations in water under pressure is thus consistent with inadequate orientational potential function dependence of the Bounds fit to the CI data points.

Is this an inadequacy in the CI data points, or the functional form chosen? The fact that the MCY parameters – albeit a poorer fit to the functional form – give a good (though not perfect) water-like g_{OO}, implies two relevant points. First, the functional form chosen is capable of giving a good approximation to water-like g_{OO} behaviour. That a good fit of this form to the CI data gives a poor water g_{OO} implies the CI dimer points are inadequate

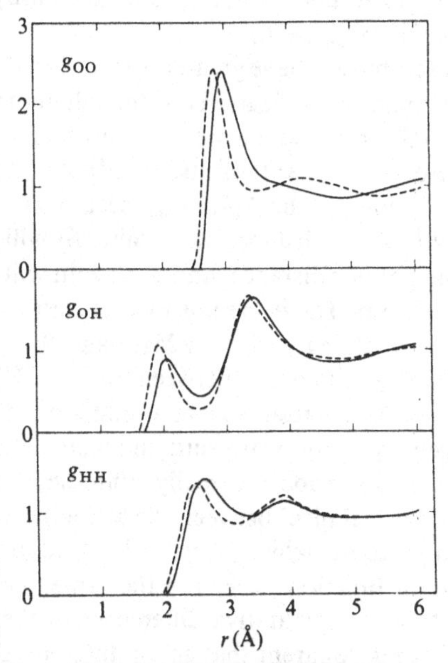

Figure 24. Partial PCFs for the Bounds' reparametrization (solid line) of the MCY potential, compared to MCY (broken line). Note the change from water-like (MCY) to simple-liquid-like (Bounds) behaviour. The shoulder on the high r side of the g_{OO} first peak is characteristic of other potentials that give simple liquid-like behaviour (e.g. PE, BF, Rowlinson); it also occurs for MCY under pressure and at elevated temperature, and thus can be related to weakened orientational correlations. Reproduced, with permission, from reference 65.

to reproduce condensed phase structure, as we should expect. Secondly, the changes in the MCY parametrization between MCY and Bounds are not large, yet the effects on g_{OO} are dramatic. This clearly shows that the computed structure is *highly* sensitive to *details* in the potential function.

To identify those parameters to which the structure is particularly sensitive is clearly of interest, but how to make this identification is not at all obvious. As figures 25(a) and (b) show, there are significant differences between the two donor and acceptor sections taken near the respective distance minima of the two potentials. Of itself, the lowering of the acceptor rotation barrier (figure 25 (a)) will not be significant in a room temperature ensemble calculation, but the related slight widening of the well in Bounds (by about 5°, see table 5) will confer some additional orientational freedom to the ensemble. The changes in the donor section are probably more significant, in particular the lowering of the *trans*→*cis* barrier by 0.6 kcal mol^{-1} will again confer additional orientational freedom on the liquid ensemble. The reason for the shift outwards of the first peak in g_{OO} (figure 24(a)) is clear from the distance plots of figure 25 (c): the minimum shifts out by about 0.1 Å. The shapes of these two sections are compared by shifting the MCY plot to coincide with the Bounds energy minimum (figure 25 (d)): both the repulsive wall and attractive regions are less steep in Bounds, leading to an increase in the 2 kcal mol^{-1} width from 1.02 to 1.09 Å (table 6).

Thus we can argue that the Bounds ensemble will have greater orientational freedom from two sources (in addition to any modifications of the inherent orientational dependence of the potential, about which we can on present evidence say little). First, the increase in r_{OO} will of itself reduce the orientational driving forces (recall in this context the $E(\theta_A, r_{OO})$ sections of figure 18). Secondly, the increased *width* of the distance well (table 6) will facilitate greater variation in the near-neighbour distance, and this again will allow the orientational structure to loosen up. The degree of this relaxation is shown in the donor and acceptor section changes (figures 25(a) and (b)).

We can thus argue that the qualitative difference between the MCY (water-like) and Bounds (simple-liquid-like) g_{OO}s results from a reduction in orientational driving forces caused largely by the outward shift and increased width of the distance well. Yet the changes made are really quite small, suggesting there is a great sensitivity in the 'balance' between the attractive/repulsive distance-dependent forces and those which specifically lead to orientational structure. As both potential function *forms* are the same, we cannot put this change in behaviour down to qualitative differences in the orientational part of the potential. It is thus apparent that small differences in the parametrization can be crucial in giving or destroying water-like behaviour (recall also the limited region of parameter space for which water-like behaviour was observed for SPC (section 4.4.1 (*a*))). This has implications for potential function design in orientation-dependent liquid systems. The constraints on obtaining *quantitative* information with which

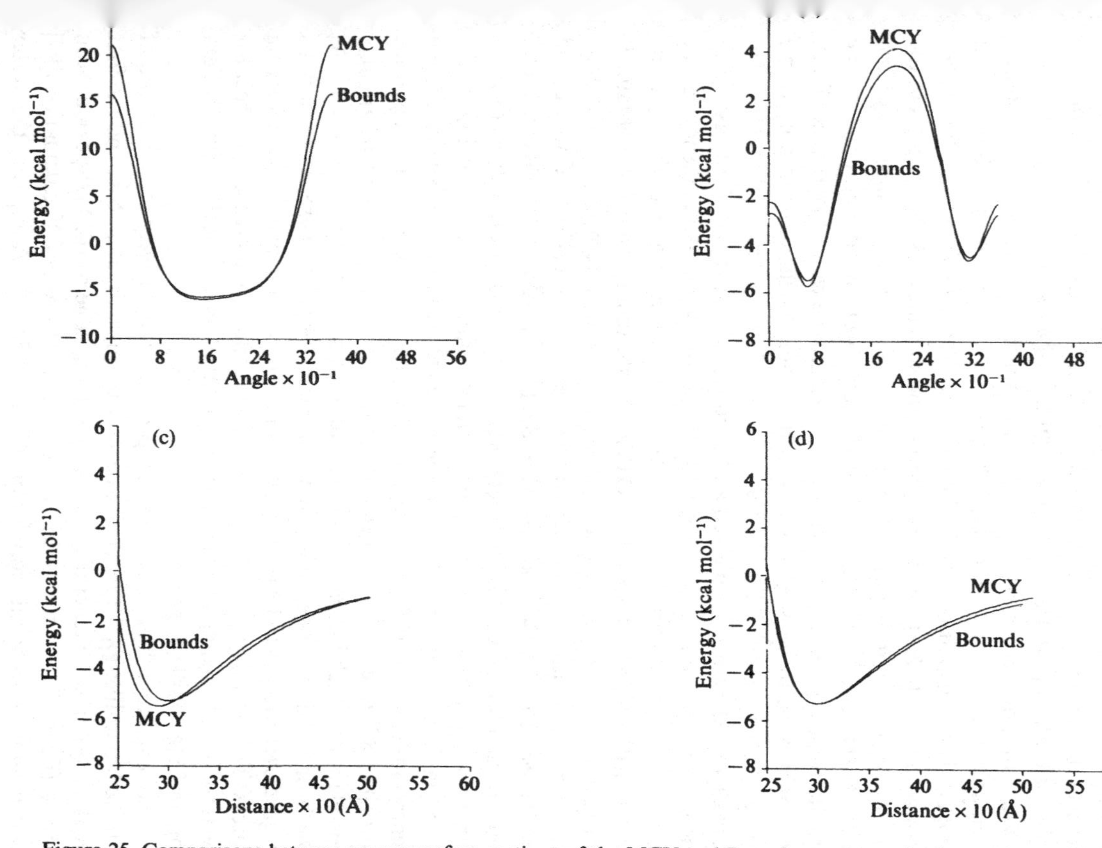

Figure 25. Comparisons between energy surface sections of the MCY and Bounds reparametrizations. (a) $E(\theta_A)$; (b) $E(\theta_D)$; (c) distance sections $E(r_{OO})$. (d) shows the distance sections with MCY shifted so that its minimum coincides with Bounds.

to parametrize a potential form are severe indeed, and unlikely to be satisfied by quantum mechanical calculations where significant quantitative uncertainties remain.

Changes in g_{OH} and g_{HH} in going from the MCY to Bounds potentials are much less dramatic than for g_{OO}, at least out to 6 Å (figure 24). The peak shifts relate to the increased O–O distance (though they are *slightly* larger, perhaps reflecting some degree of increased orientational freedom). The only apparently significant differences relate to the breadths of the first peak in each case. These can be related to increased orientational freedom in the Bounds case, but we note that they are quantitative rather than qualitative changes, underlining earlier comments (section 2) concerning relative insensitivity of both g_{OH} and g_{HH} as structural informants.

We can not yet assert that accurate dimer potentials are inherently incapable of giving water-like behaviour, as is often done when cooperative/non-pair-additive effects are invoked to explain away failures of proposed potentials; though considering the smallness of the changes in the MCY case, one might argue that the expected additional binding energy from such effects could reduce the potential-energy well breadth to enhance the orientational correlations sufficiently to pass into the water-like regime. Thus, we might hope to be able to see such transitional behaviour in polarizable models, run with and without polarizability. Although this could not be seen in earlier, prematurely-truncated and hard repulsive core (r^{-12}) forms of PE, there are suggestions in recent work[**43**] that the water-like tendency can be increased by certain modifications. An r^{-12} repulsive core is both physically unjustified, and suppresses the O–O distance shrinkage that can be obtained; softening (to $r^{-7.3}$) gives a much better fit to the quantum mechanical data. The dimer separation, (contracted) O–O distance in ice-Ih, and the second virial coefficient can then be used to fit the Lennard–Jones (7:3:6) constants; only a small region of this parameter space gives reasonable values for these three sets of fitting data. With these modifications some tightening of the potential in the right direction can be obtained. Adding repulsive hydrogen centres (as exist in MCY, and suggested important by recent crystallographic work[**88**, **89**] (see section 7 below)) results in an enhancement of angular correlations which increases the population of pairs at approximately 4.5 Å; the broad trailing edge of the first g_{OO} peak (obtained here, as in Bounds (figure 24)) appears to be being driven out by increasing orientational driving forces. This addition of hydrogen–hydrogen repulsion increases the acceptor section barrier heights closer to the quantum mechanical reference data (figure 21(a), table 5, section 5.6).

The remaining dimer potentials all produce qualitatively the same kind of simple liquid-like g_{OO}: the second peak is at approximately 6 Å, and the first peak is too broad. Generally, there is a trailing edge to the first peak as seen above (figure 24) in the Bounds and high pressure MCY calculations[**90**] which show significant population of the 3.25–3.75 Å region; the water molecules that should be kept out at approximately 4.5 Å by the strength of the orientational dependence of the potential function appear to drift in. The orientational driving forces are too weak. Coordination numbers measured to the first minimum are larger than in water-like systems. The heights of the first peaks of these model 'waters' also vary: HF and Rowlinson are good (~ 2.3–2.5), while Watts (3.55) and BF (2.71) are too high (cf. table 7). PE can be too high or about right, depending upon the version of the model used.

The above discussion on the MCY/Bounds potentials suggest we look in a little more detail at the effects of adding correlation terms onto the HF potential. Referring to table 6, we see that the HF well is very wide: moreover, as we see from the quantum mechanical results quoted, taking the quantum mechanical calculation to CI level reduced the well width very significantly. We would thus expect the addition of a dispersion term to an SCF-derived potential to enhance the orientational correlations, by pulling in the first neighbour molecules, and so increase the 'water-like' properties of the PCF. We can examine the effect of adding correlation effects by considering the HF modifications made by Kistenmacher *et al.* **[59]** and their effect on the g_{OO} PCF.

We consider the effects on the g_{OO} PCF from the inclusion of two dispersion corrections chosen to cover the maximum range expected. These are the London and Kirkwood–Müller corrections discussed in section 4, which increase the $E(r_{OO})$ well *depth* by about 1 and 2 kcal mol^{-1} respectively in our standard linear geometry (table 6). The minimum O–O separation is also shifted to lower r, as expected. Moreover, the potential wells are narrowed, from about 1.28 Å (HF) to about 1.02 Å (HFL) and 0.92 Å (HFK); the HFL width is about the same as in the MCY potential. As the underlying form is similar, we might expect a significant shift to water-like behaviour as the narrower distance well tightens up the orientational driving forces.

What we in fact observe is slightly strange. From the MCY/Bounds discussion we would expect (a) an increase in the height of the first peak, associated with a well narrowing, and (b) the appearance of an approximate 4.5 Å peak consequent upon the shifting of neighbours from the trailing edge of the originally broad first peak (perhaps with some shift inwards to approximately 4.5 Å from the 6 Å peak). This in fact we observe. However, we would also expect the effect to be greater for HFK (the narrower well) than HFL, whereas the published trend is in the other direction (figure 26) – HFK has a weaker second peak, and also retains a trailing edge to the first. This discrepancy is so odd that one is tempted to look further, and examination of the Γ_{OO} curves ($\Gamma(r) \equiv 2\pi\rho r^2 g(r) dr$) (figure 26) suggests an inconsistency between the g_{OO} and Γ_{OO} plots for both HKF and HFL in the original paper. This inconsistency can be removed by a simple relabelling of the g_{OO} curve HFK to HFL and vice versa. If we reconsider the curves in the light of this transposition, the pattern fits: narrower wells strengthen the orientational correlations to enhance the water-like behaviour of the liquid.

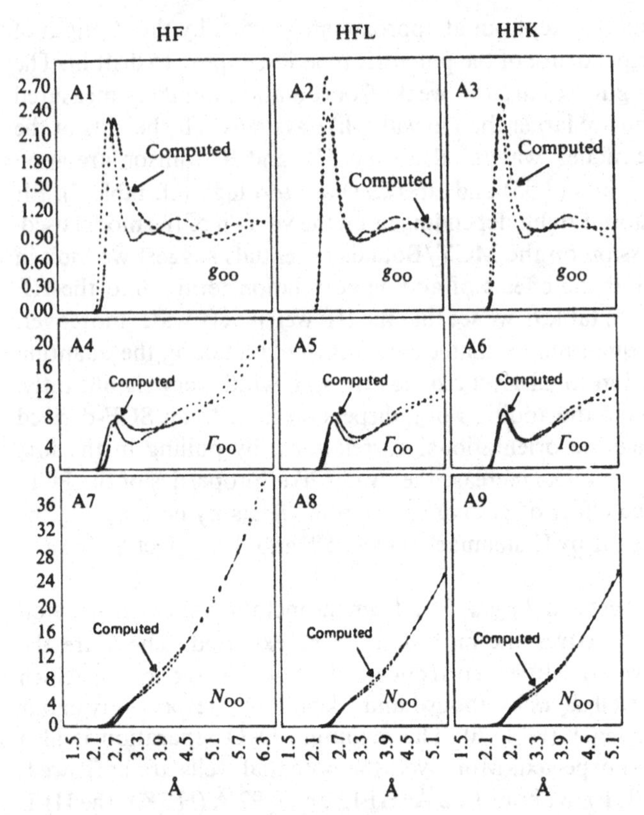

Figure 26. $g_{OO}(r)$ (top), $\Gamma_{OO}(r)$ (middle) and $N_{OO}(r)$ (bottom) for the HF, HFL and HFK potentials from reference 59. $\Gamma(r)[\equiv 2\pi r^2 \rho g(r) dr]$ gives the number of particles in a spherical shell of radius r, while $N(r)[\equiv \int_0^r \Gamma(r) dr]$ is the cumulative coordination number. A2 and A3 seem inconsistent with A5 and A6 respectively. Exchanging A2 and A3 would remove this inconsistency, and also make sense with respect to the different depths of the potential function wells (figure 12). Reproduced, with permission, from reference 59.

This orientational structure as evidenced by the second peak is still insufficient, and the first peak too high, implying the distance well is probably too narrow and the orientational energy dependence of the underlying form too weak. It suggests that enhancing the orientational structure by narrowing the distance well is done at the expense of raising h_1, the first peak in g_{OO}. To obtain adequate orientational structure *without* too high an h_1 implies that we need a *broader* distance well, and some compensating adjustment to the underlying orientational dependence of the potential function. We return to this point in the concluding section.

Before leaving the HF potential and its relatives, we should comment on the fact that adding the very strong Kirkwood–Müller dispersion term does not enhance the water-likeness of the liquid to the extent found in MCY, yet HFK has a much narrower distance well. Moreover, the underlying form is the same in both cases, apart from the way in which the dispersion terms are included (they are *anisotropic* in MCY, isotropic in the other cases). Presumably the difference must relate to a relative weakness in the orientational dependence of the HF potential resulting from the different parametrization and/or the difference in the way the dispersion term is included. On currently-available evidence, we can say nothing about the first possibility. The dispersion terms are, however, of different forms. In the HFK/L cases, the classical $-C_6r^{-6}$ term used is centred on the oxygen, and it is thus spherically symmetrical. For MCY, in contrast, the form is more complex, being an inverse exponential form between each intermolecular O–H pair; it is thus significantly anisotropic, as well as having a different explicit distance dependence. As the HFK and HFL g_{OO} PCFs implied insufficiently strong orientational correlations, the non-central dispersion term in MCY may be the cause of orientational enhancement, though at present this is only a little more than pure speculation: reference to figure 20 shows an increase from 1.8 to 2.6 kcal mol^{-1} in the donor *trans*→*cis* barrier, implying the HF→MCY changes are not isotropic to the order of kt and hence changes in orientational dependence of the potential may be significant.

It remains to comment on the remaining 'true dimer' potentials of table 6 that give water-like g_{OO} PCFs. QPEN we consider no further, as it was parametrized using MCY; we observe only that the resulting g_{OO} has a sharper and narrower first peak than MCY, presumably reflecting the influence of the QPEN form in imposing a stronger orientational dependence. That STO-3G gives a water-like structure is not surprising, as it is well-known that such small basis set quantum mechanical calculations overestimate the intermolecular attraction. This was clear following the early pioneering work of Del Bene and Pople.[87] The first peak in g_{OO} is both too high (~ 3.4) and too narrow, consistent with the very narrow distance well (~ 0.57 Å) (table 6). The second peak is also too strong. Both peaks are at too short distances (2.70 and 4.47 Å), reflecting the underestimation of the equilibrium dimer distance resulting from the overstrong intermolecular interaction.

The Clementi–Caravetta (CC) and RWK1 potentials are more problematical. In the CC case, the strength of the dispersion term results in a narrow distance well (~ 0.90 Å). The strength of the orientational dependence of the potential energy of the underlying Clementi form seems to be altered by reducing (with respect to MCY) the charge (by 9 per cent) and the distance between the oxygen and the negative charge centre, though the effect on the dipole is an increase of only a few per cent. The effect on the donor and acceptor surfaces is to widen slightly the $E(\theta_A)$ well, (table 5) without affecting significantly the *trans*→*cis* barrier in the donor section (figure 20); thus there is a small apparent relaxation of orientational restriction. The effect of the dispersion term on the distance $E(r_{OO})$ well is large, as expected from the 13 per cent width reduction (table 6). The effect is to make steeper both the repulsive wall and the attractive regions of the curve. The overall consequence of these alterations for the g_{OO} PCF are not surprising: water-like behaviour is observed (figure 4 of reference 67), though the second peak is insufficiently strong. Despite the strong distance potential (a narrow distance well), the underlying orientational correlations in the potential appear too weak to be sufficiently enhanceable by the strong distance-dependent attraction. This in turn results in too high a first peak, and hence too large a coordination number, by now a familiar effect. It is however interesting that the first peak of g_{OO} is perhaps too broad rather than too narrow, despite the narrow distance well.

The RWK1 result is not straightforward to rationalize. The original Watts model was a dimer potential fitted to gas phase data, which gave a normal simple liquid-like g_{OO}, in common with most of the other 'true dimer' potentials. RWK1 was derived by shifting the centre of negative charge from the oxygen along the symmetry axis, in the manner of the other displaced charge models (e.g. HF, MCY, TIPS2, BF), fitting the shift δ and charge q to isolated molecule dipole and quadrupole values. These changes result in a water-like g_{OO} with a too-weak second peak at 4.5 Å, and a very strong (4.00) and narrow first peak (table 7). We might note that the dispersion term here included C_8r^{-8} and $C_{10}r^{-10}$ terms, which would act to steepen the attractive region of the potential and hence enhance the inherent orientational dependence of the potential. The use of the softer inverse exponential repulsion would serve to increase this effect over that of a harder repulsive wall (r^{-12}) with the same 2 kcal mol^{-1} well width. It is of interest that changing the dispersion correction as was done in RWK2 caused a reversion to simple liquid-like behaviour; as seven repulsive parameters had in consequence to be refitted (from second virial and ice data), it is unclear as to the major reason for this reversion, although the acceptor well does seem to be rather wide (table 5). If we accept that the dispersion term in RWK2 is more satisfactory, then we still seem to have a problem with the orientational dependence of the potential. Using this same (spherically symmetrical) dispersion term in conjunction with HF was insufficient to give a recognizable water-like g_{OO}.[61]

7. Some Tentative Conclusions

We now attempt to draw together some conclusions from the above discussion, and suggest ways in which improvements might be made to potentials within the framework of rigid body, classical models.

No current model is satisfactorily able to reproduce *quantitatively* the structure of water. As argued previously, a water-like g_{OO} is not enough: quantitative agreement with experimental positions, heights and areas of the g_{OO} peaks is necessary, and no model yet proposed satisfies this condition even at room temperature and pressure.

Most of the models considered show similar acceptor and donor angle dependencies. There are, however, quantitative differences in barrier heights and well widths which seem to fit a pattern. The four-point charge models have stronger angular constraints, with a significant barrier in the donor section separating the *cis* and *trans* dimers; it is about the same height as implied by the quantum mechanical reference data. The acceptor angle sections are narrower than expected from quantum mechanics, and are distinctive in having a definite barrier between *cis* and *trans* configurations in this section also: we suspect this *strong* orientation dependence of the potential is the major reason that these models have little problem in producing a water-like g_{OO} though if anything, the $1.6r_0$ peak tends to be too strong. The first peak is also too high and too sharp.

The three-point-charge (both three and four centre) models are interesting; they seem to be less stable, in that small changes in model parameters can lead to a loss of apparent water-like behaviour, with a consequent reversion to a simple liquid-like g_{OO}. The *cis/trans* separation in the 4/5-type acceptor section is lost, the resulting well shape and width for the 3/4 models being in good agreement with the quantum mechanical data. In contrast, the 3/3 models fail to reproduce the near-linear behaviour in the bottom of the well (figure 21), and have much narrower acceptor sections (table 5). For both 3/3 and 3/4 types, there are examples of parametrizations which fail to yield water-like behaviour, and here again a pattern emerges which suggests why. Considering first the 3/3 models, TIPS has a much lower *trans*→*cis* barrier, and a much shallower global well, than SPC (figure 20); it is tempting to assign its failure to give a water-like g_{OO} to this lower orientation dependence of the potential. The TIP3P reparametrization both deepens the global minimum and increases the *trans*→*cis* barrier to the same order as that of SPC. There is also a slight reduction in the acceptor well width (table 5). A large reduction in the width of the distance well also occurs. Thus, in inducing (or losing) water-like behaviour, we have a pattern similar to that observed in going from Bounds to MCY (or *vice versa*). What appears to be happening in losing water-like behaviour is that, as the effective orientational driving forces are weakened (TIP3P→TIPS; MCY→Bounds), there comes a point at which the extended water-like network structure is unable to support itself, and a catastrophic collapse occurs to a simple liquid-like structure with much weaker orientational correlations. We might even draw a tentative analogy to a phase transition, the two types of structure being the two 'phases'.

A similar, though quantitatively less clear pattern is found with the 3/4 models. Again, there are differences in the donor section *trans*→*cis* barrier heights (figure 20), with the lowest one being given by the model (BF) which gives a simple-liquid-like g_{OO}. The differences in the acceptor section are, however, larger (table 5) with both BF and RWK2 – both of which fail to give clear water-like behaviour – having wider wells. There are no clear differences in the distance well widths (table 6).

For all three forms which can yield either water-like or simple-liquid-like behaviour, we thus have a similar pattern: there seems to be a degree of orientational dependence of the potential function below which a water-like network structure cannot be supported. In some cases (e.g. the 3/3 models), this change in orientational dependence may relate, at least partially, to an enhanced distance attraction, whereas in others (the 3/4 models), the changes seem to be more in the orientational dependence itself. The MCY-type potentials seem to show a mixture of such behaviour. In all cases, the changes that are needed to gain or lose water-like behaviour are small, and may relate to different aspects of the energy surface in different cases, underlining a high sensitivity to the parametrization. We might also generalize that the more complex the inherent orientational dependence of a model, the greater the tolerance to wider acceptor angle wells (cf. TIPS2 and TIPS, table 5) and to lower donor section *trans→cis* barriers (cf. MCY and TIPS, figure 20). Of all the models, only the 3/4 and MCY-types do not have artificially narrow acceptor wells (table 5). These seem to be characteristics worth retaining.

The above discussion suggests that we might usefully consider a potential function in terms of (i) its *underlying* orientational dependence and (ii) its distance dependence. Although such a separation may appear artificial, it is conceptually helpful. For example, for a given underlying orientational dependence, we could enhance the expected orientational correlations in the liquid by the simple device of narrowing the distance well by increasing the strength of the attractive part: we thereby draw in the neighbours, and force out some of the intermolecular orientational freedom. This is in effect what we are doing in adding dispersion terms to HF, or moving from the Bounds to MCY models. However, this procedure appears not a totally satisfactory way of improving a water potential: it invariably results in (a) too high a first g_{OO} peak (h_1) and (b) an inadequate second g_{OO} peak (h_2) relative to h_1. The first effect implies *the enhanced attraction is too strong*, a conclusion consistent with the resultant distance wells being narrower than expected from the quantum mechanical calculations. The second problem implies that *the underlying orientational dependence of the potential is too weak*, a conclusion again consistent with the known too-rapid breakdown in water-like behaviour in MCY calculations at increased temperature and pressure.

The characteristics of the three-point charge EPPs are consistent with this hypothesis: h_1 is too strong and h_2 too weak. This implies the distance attraction is too strong (and the EPP distance wells are all *much* too narrow compared to the quantum mechanical reference data), and the underlying orientational dependence too weak. Thus, as enhancing the liquid orientational structure by strengthening the distance well artificially enhances the first peak in g_{OO}, *it suggests again that we need rather to weaken* $E(r_{OO})$ *and strengthen up the orientational dependence somewhat.*

We have two examples of potentials in which the orientational dependence is stronger, namely BNS and ST2. Both also have narrow distance wells, and are generally expected also to have too strong orientational energy dependence.

Interestingly, BNS has a second peak which is if anything too high (see h_1/h_2 ratios in table 7), and this is the potential with the strongest orientation dependence; reducing the strength of the distance well would be expected to reduce h_1 to something closer to experiment, as would reducing the orientation dependence. The latter was done in developing ST2, but the strong distance dependence resulted in far too high a first peak in g_{OO}. It would seem that we need to reduce the distance-dependent attraction, and modify the orientation dependence in some other way to avoid any possible resulting over-loss of orientational structure in the simulated liquid.

We thus seem to have a consistent hypothesis: for most existing potentials that give water-like behaviour, the underlying orientational dependence of the potential is too weak (or perhaps incorrect in some section we have not considered) while the distance dependence is too strong. Enhancing orientational structure in the liquid by increasing the strength of the radial term results in too strong a first peak in g_{OO}. Further progress seems to require (a) a reduction in the narrowness of the distance well and (b) modifications (probably a strengthening) in the underlying orientational dependence in some region of the surface.

How to go about such modifications is not clear. The donor and acceptor angle dependencies of most water-like models are similar in form both between themselves and to the quantum mechanical reference data; inducing differences in wells and barriers by different parametrizations seems unable to decouple h_1 and h_2. Although it is perhaps too easy to invoke non-pair-additive effects as a way out, the possibility merits serious consideration, remembering the significant change that polarizability can make to the relative contributions of dipole–dipole, dipole–quadrupole and quadrupole–quadrupole energies and their associated structural tendencies (section 3.6). An interesting indication of the possibility of non-pair-additive effects strengthening orientation and distance *correlations* in a simulated liquid comes from liquid simulations of PE(QQ) with and without polarizability. **[43]** As polarizability is switched on, not only is there a tightening up of the geometry of the first-neighbour hydrogen bond, but also an increase in the population of second neighbours at about 4.6 Å at the expense of a population reduction at shorter distances; the high r shoulder observed on the first peak of the simple liquid-like g_{OO} (e.g. figure 24) appears to be beginning to shift out to longer distances as the orientational correlations tighten up. This effect seems to be similar to that observed previously on adding in hydrogen centres and softening the repulsive core (section 6.2).

This 'hydrogen-bond tightening' from switching on polarizability would be expected to increase the height (h_1) and sharpness of the first g_{OO} peak. The effect thus seems to be just that which results from narrowing the distance well, in that enhancing the population at about 1.6 r_0 is done at the expense of increasing h_1. A direct indication that the effect of many-body terms parallels that of narrowing the distance well comes from very recently-published work of Detrich *et al.* [**93**], in which MCY simulations are run with the addition of three- and four-body terms. The effects are to increase both h_1 and h_2 by about the same relative amounts, with only a very small (perhaps not significant?) reduction in the h_1/h_2 ratio from 2.28 to 2.22, still higher than the experimental value of about 2.1. There is also a very considerable narrowing of the first peak, as we also find when we reduce the width of the distance well of other potentials. It would appear therefore that we need to look elsewhere for potential function modifications that will break this linking of h_1 and h_2.

One possibility of orientational modification is suggested by both recent experimental and theoretical work in related areas. Most of the formalisms discussed above which yield water-like potentials use a single-centre isotropic repulsion. An exception is MCY, which includes OO, OH and HH repulsions; interestingly, in terms of h_1 and the h_1/h_2 ratio (table 7), this potential gives perhaps the best agreement with experiment of all so far considered (though there are not insignificant problems with the position of the second peak). Recent high-resolution crystallographic studies on vitamin B_{12} coenzyme crystals have resulted in a well characterized solvent network in which the hydrogen atoms of many of the waters have been located. [**8**, **88**, **89**] Two results of this study are of particular interest. First, the relative water molecule orientations of a chain of solvent molecules can most simply be rationalized in terms of next-neighbour H–H repulsions, the HH separation distances being all approximately equal to the accepted van der Waals radius sum. It thus appears that the final water molecule orientations are being controlled by the proton–proton repulsions, rather than the orientational dependence of the electrostatic part of the potential. Similar regularities have been found by Savage [**88**, **89**] in several other water-containing structures, tempting a conclusion that H–H repulsions may be a major factor controlling water structure. McDonald and Klein [**91**] also noted a correlation between simple liquid-like behaviour and poor H–H repulsion. Preliminary studies by Quinn on a modification of PE($\mu\Omega$) including hydrogen repulsions shows both enhanced orientational correlations leading to an increased second neighbour population in the $1.6r_0$ region of g_{OO}, together with a significant tightening up of first neighbour orientations. There are thus at least initial indications that the inclusion of specific H–H repulsion is leading to increased water-like behaviour.

A second conclusion by Savage (personal communication) concerns a significant anisotropy in the repulsive core around the lone-pair region, which it is also necessary to invoke to explain the orientational structure of the water molecules in the B_{12} crystal. This latter suggestion is also consistent with theoretical work. For example, Brobjer and Murrell [**41**] required an anisotropic overlap repulsion to fit their quantum mechanical calculations on HF (section 4.3), and similar conclusions emerge from current work on potential function design from the Cambridge group.[92] Applying such anisotropies to the water–water interaction is in its infancy, but its seems a valid direction for development. We would be decreasing the strength of the distance-dependent attraction as we proposed was necessary above, but also simultaneously increasing the orientational dependence of the potential, which also seems to be necessary. We believe that future progress is most likely in this direction.

The forms of the overlap repulsion and dispersion attraction contributions also require comment. First, there seems no justification for persevering with inverse twelfth power repulsions: these are theoretically far less justifiable than an exponential form, and appear to be far too steep (so steep, in fact, that they prevent any significant distance contraction from non-pair-additive effects in polarizable models). Reducing the steepness may be a way of increasing the width of the distance well, and *at the same time* increasing the strength of the orientational part of the potential function: we might recall that the most successful potential (MCY) has a distance well which is not excessively narrow compared to the quantum mechanical reference data (table 6), and also has an exponential repulsive form. Secondly, the form and parameters chosen for the dispersion attraction are critical in determining the shape of the attractive part of the distance well. Certainly there is no justification for ignoring the term altogether, and there are suggestions from the Watts-derived potentials that including C_8 and C_{10} terms may be important. Certainly, this will increase the steepness of the attractive side of the distance well, and hence increase, at least slightly, the effects of the orientational part of the potential. The relatively greater success of potentials with anisotropic dispersion terms also suggests there may be a case for such an anisotropy, as is included in MCY and its derivatives (see discussion in section 6.2).

Finally, we would underline two points we consider particularly important. First, comparisons between experimental and computed PCFs must be made critically: it is not enough to obtain a water-like g_{OO} – there must be a quantitative agreement to ensure important structural detail is reproduced by the model. Secondly, the liquid structure resulting from a particular potential function appears to be the result of a delicate balance of competing effects, which we have tried to rationalize in terms of the orientation and distance-dependent parts of the potential. *Small* changes in the parameters assigned to a potential form may result in *qualitative* changes in g_{OO}, for example in going from MCY to Bounds (section 6.2), or in optimizing SPC (section 4.4.1 (*f*)). This illustrates the sensitivity of simulated liquid water structure to chosen parameters, and underlines the problems in devising potential functions that are quantitatively adequate. The phase transition analogy between simple liquid-like and water-like behaviour suggested earlier may also be a useful one to follow up.

Acknowledgements

Many of the ideas developed here have arisen through arguments over several years in the Birkbeck Liquid Group, involving especially Paul Barnes, Barry Gellatly and Julia Goodfellow. Discussions with David Beveridge, Mihail Mezei, Vic Saunders, Sarah Price, Bruce Berne and Peter Kollman are particularly acknowledged. Hugh Savage is responsible for the ideas derived from the coenzyme B_{12} work. We recognize the speculative nature of some of the diagnoses; the responsibility is J.L.Fs. J.O.B. thanks the SERC for a Research Studentship.

Note. J. E. Quinn's current address is: DAP Support Unit, Queen Mary College, Mile End Road, London E1 UK.

References

1. J. D. Bernal. *Proc. Roy. Soc.* A **280** (1964), 299–322.
2. J. L. Finney. *Proc. Roy. Soc.* A **319** (1970), 479–93.
3. J. L. Finney. *Proc. Roy. Soc.* A **319** (1970), 495–507.
4. J. M. Ziman. *Models of Disorder*. Cambridge University Press, Cambridge, 1979, p. 78.
5. J. A. Pople. *Proc. Roy. Soc.* A **205** (1951), 163–78.
6. J. D. Bernal & R. H. Fowler. *J. Chem. Phys.* **1** (1933), 515–48.
7. J. L. Finney, B. J. Gellatly, I. C. Golton & J. Goodfellow. *Biophys. J.* **32** (1980), 17–30.
8. J. L. Finney. *J. de Phys.* **45** (Colloque C7, suppl. 9) (1984), C7197–209.
9. J. C. Dore (this volume).
10. C. H. Bennett. *J. Appl. Phys.* **43** (1972), 2727–34.
11. T. R. Dyke, K. M. Mack & J. S. Muenter. *J. Chem. Phys.* **66** (1977), 498–510.
12. A. H. Narten, M. D. Danford & H. A. Levy. *Disc. Farad. Soc.* **43** (1967), 97–107.
13. J. P. Hansen & I. R. McDonald. *Theory of Simple Liquids*. Academic Press, New York, 1976.
14. R. W. Watts & I. J. McGee. *Liquid State Chemical Physics*. Wiley-Interscience, New York, 1976.
15. C. A. Croxton. *Liquid State Physics*. Cambridge University Press, Cambridge, 1974.
16. J. P. Valleau & S. G. Whittington. In *Modern Theoretical Chemistry*, Vol. 5 (ed. B. J. Berne). (*Statistical Mechanics, Part A: Equilibrium Techniques*) Plenum Press: New York, 1977, p. 137.
17. D. W. Wood. In *Water – A Comprehensive Treatise*, Vol. 6 (ed. F. Franks), Plenum Press: New York, 1979, p. 279.
18. A. Ben-Naim. *Water and Aqueous Solutions*. Plenum Press, New York, 1974.
19. H. C. Andersen. *J. Chem. Phys.* **59** (1973), 4714–25.
20. H. C. Andersen. *J. Chem. Phys.* **61** (1974), 4985–92.
21. L. W. Dahl & H. C. Andersen. *J. Chem. Phys.* **78** (1983), 1980–93.
22. C. A. Angell. In *Water – A Comprehensive Treatise*, Vol. 7 (ed. F. Franks). Plenum Press: New York, 1982, p. 1.

23. R. M. Pettitt & P. J. Rossky. *J. Chem. Phys.* **77** (1982), 1451–7.
24. D. Chandler & H. C. Andersen. *J. Chem. Phys.* **57** (1972), 1930–7.
25. H. J. C. Berendsen, J. P. M. Postma, W. F. van Gunsteren & J. Hermans. In *Intermolecular Forces* (ed. B. Pullman). Reidel: Dordrecht, 1981, p. 331.
26. J. E. Enderby & G. W. Neilson. *Adv. in Phys.* **29** (1980), 323–65.
27. G. N. Patey & J. P. Valleau. *J. Chem. Phys.* (1976), 170–84.
28. P. Barnes. In *Report of CECAM Workshop on Molecular Dynamics and Monte Carlo Calculations on Water* (ed. H. J. C. Berendsen). CECAM: Paris, 1972, p. 77.
29. P. Barnes, J. L. Finney, J. D. Nicholas & J. E. Quinn. *Nature* **282** (1979), 459–64.
30. D. Hankins, J. W. Moskowitz & F. H. Stillinger. *J. Chem. Phys.* **53** (1970), 4544–54.
31. E. Clementi, W. Kołos, G. C. Lie & G. Ranghino. *Int. J. Quant. Chem.* xvii (1980), 377–98.
32. B. J. Gellatly, J. E. Quinn, P. Barnes & J. L. Finney. *Mol. Phys.* **59** (1983), 949–70.
33. E. S. Campbell & M. Mezei. *J. Chem. Phys.* **67** (1977), 2338–44.
34. P. Barnes. In *Progress in Liquid Physics* (ed. C. A. Croxton). Wiley-Interscience: Chichester, 1978, p. 391.
35. H. J. C. Berendsen & G. A. van der Velde. In *Report of CECAM Workshop on Molecular Dynamics and Monte Carlo Calculations on Water* (ed. H. J. C. Berendsen). CECAM: Paris, 1972, p. 63.
36. F. H. Stillinger & C. W. David. *J. Chem. Phys.* **69** (1978), 1473–84.
37. J. M. Goodfellow. *Proc. Nat. Acad. Sci. USA* **79** (1982), 4977–9.
38. F. H. Stillinger. *J. Phys. Chem.* **74** (1970), 3677–87.
39. F. H. Stillinger. *J. Chem. Phys.* **57** (1972), 1780–7.
40. F. H. Stillinger. *Adv. Chem. Phys.* **31** (1975), 1–101.
41. J. T. Brobjer & J. N. Murrell. *Mol. Phys.* **50** (1983), 885–99.
42. A. J. Stone. *Mol. Phys.* **43** (1981), 233–48.
43. J. E. Quinn. Unpublished work.
44. F. London. *Z. Phys. Chemie* **B 11** (1930), 222–51.
45. R. Mecke & W. Baumann. *Phys. Zeits.* **33** (1932), 833–5.
46. J. S. Rowlinson. *Trans. Farad. Soc.* **45** (1949), 974–84.
47. J. S. Rowlinson. *Trans. Farad. Soc.* **47** (1951), 120–9.
48. A. Rahman & F. H. Stillinger. *J. Chem. Phys.* **55** (1971), 3336–59.
49. A. Ben-Naim & F. H. Stillinger. In *Structure and Transport Processes in Water and Aqueous Solutions* (ed. R. A. Horne). Wiley-Interscience: New York, 1972, p. 295.
50. F. H. Stillinger & A. Rahman. *J. Chem. Phys.* **60** (1974), 1545–57.
51. O. Weres & S. A. Rice. *J. Amer. Chem. Soc.* **94** (1972), 8983–9002.
52. W. L. Jorgensen. *J. Amer. Chem. Soc.* **101** (1979), 2011–15.
53. L. L. Shipman & H. A. Scheraga. *J. Phys. Chem.* **78** (1974), 909–16.
54. O. Matsuoka, E. Clementi & M. Yoshimine. *J. Chem. Phys.* **64** (1976), 1351–61.
55. F. T. Marchese, P. K. Mehrotra & D. L. Beveridge. *J. Phys. Chem.* **85** (1981), 1–3.
56. E. Clementi & H. Popkie. *J. Chem Phys.* **57** (1972), 1077–94.
57. H. Popkie, H. Kistenmacher & E. Clementi. *J. Chem. Phys.* **59** (1973), 1325–36.
58. H. Kistenmacher, G. C. Lie, H. Popkie & E. Clementi. *J. Chem. Phys.* **61** (1974), 546–61.

59. H. Kistenmacher, H. Popkie, E. Clementi & R. O. Watts. *J. Chem Phys.* **60** (1974), 4455–65.
60. D. Eisenberg & W. Kauzmann. *Structure and Properties of Water*. Oxford University Press London, 1969.
61. J. R. Reimers, R. O. Watts & M. L. Klein. *Chemical Physics* **64** (1982), 95–114.
62. C. Douketis, G. Scoles, S. Marchetti, A. J. Thakkar & M. Zen. *J. Chem. Phys.* **76** (1982), 3051–63.
63. J. Hepburn, G. Scoles & R. Penco. *Chem. Phys. Lett.* **36** (1975), 451–56.
64. R. Allricks, R. Penco & G. Scoles. *Chem. Phys.* **19** (1977), 119–30.
65. D. G. Bounds. *Chem. Phys. Lett.* **96** (1983), 604–6.
66. E. Clementi & P. Habitz. *J. Phys. Chem.* **87** (1983), 2815–20.
67. V. Carravetta & E. Clementi. *J. Chem. Phys.* **81** (1984), 2646–51.
68. R. Colle & O. Salvetti. *Theor. Chim. Acta* **37** (1975), 329–44.
69. R. Colle & O. Salvetti. *Theor. Chim. Acta* **53** (1979), 55–63.
70. W. L. Jorgensen. *J. Amer. Chem. Soc.* **103** (1981), 335–40.
71. W. L. Jorgensen. *J. Chem. Phys.* **77** (1982), 4145–63.
72. W. L. Jorgensen, J. Chandrasekhar, J. D. Madura, R. W. Impey & M. L. Klein. *J. Chem. Phys.* **79** (1983), 926–35.
73. R. O. Watts. *Chem. Phys.* **26** (1977), 367–77.
74. H. L. Lemberg & F. H. Stillinger. *J. Chem. Phys.* **62** (1975), 1677–90.
75. A. Rahman, F. H. Stillinger & H. L. Lemberg. *J. Chem. Phys.* **63** (1975), 5223–30.
76. A. D. J. Haymet & D. W. Oxtoby. *J. Chem. Phys.* **77** (1982), 2466–74.
77. B. R. Lentz & H. A. Scheraga. *J. Chem. Phys.* **58** (1973), 5296–308.
78. G. H. F. Diercksen. *Theor. Chim. Acta* **21** (1971), 335–67.
79. J. O. Baum & J. L. Finney. *Mol. Phys.* In press.
80. J. O. Baum, PhD thesis, University of London, 1985.
81. S. F. Boys & F. Bernardi. *Mol. Phys.* **19** (1970), 553–66.
82. V. R. Saunders. Personal communication.
83. S. R. Langhoff & E. R. Davidson. *Int. J. Quant. Chem.* **8** (1974), 61–72.
84. M. Mezei & D. L. Beveridge. *J. Chem. Phys.* **76** (1982), 593–600.
85. M. D. Newton & N. R. Kestner. *Chem. Phys. Lett.* **94** (1983), 198–201.
86. J. C. Dore. *Farad. Disc. Chem. Soc.* **66** (1978), 82–3.
87. J. Del Bene & J. A. Pople. *J. Chem. Phys.* **52** (1970), 4758–66.
88. H. F. J. Savage, PhD thesis, University of London, 1983.
89. H. F. J. Savage. In preparation.
90. R. W. Impey, M. L. Klein & I. R. McDonald. *J. Chem. Phys.* **74**, 647–52.
91. I. R. McDonald & M. Klein. *Farad. Disc. Chem. Soc.* **66** (1978), 48–57.
92. S. L. Price & A. J. Stone. *Mol. Phys.* **40** (1980), 805–22.
93. J. Detrich, G. Corongui & E. Clementi. *Chem. Phys. Lett.* **122** (1984), 426–30.

Alcohol–Water Mixtures Revisited

FELIX FRANKS* AND JACQUES E. DESNOYERS†

**Department of Botany, University of Cambridge, Cambridge, CB2 3EA, U.K.;*
†Institut National de la Recherche Scientifique, Ste Foy, Quebec G1V 4C7, Canada

1. Historical Introduction

Aqueous solutions of alcohols have been for many years, and are still today, of considerable interest to a wide range of scientists and technologists. A low cost and miscibility with water of the lower members of the homologous series renders such mixtures useful as industrial solvent media for a variety of chemical reactions and for small and large scale separation processes. In particular, aqueous solutions of alcohols are often employed in the extraction and manipulation of labile materials such as proteins.

Physical chemists have long been intrigued by the eccentric properties of the mixtures, especially in the low concentration range. A review, published in 1966, and entitled 'The structural properties of alcohol–water mixtures' highlighted and tried to explain these peculiar features in the light of the then current ideas about the intermolecular nature of liquid water and solute–solvent hydrogen bonding.[1] Emphasis was on thermodynamic properties with some excursions into spectroscopic, dielectric and transport phenomena. The general conclusion was that the predominant interaction, at least in dilute solutions, is not hydrogen bonding between the two species, but is determined primarily by the alkyl residue. The review posed more questions than it was able to answer but it stimulated considerable interest and activity among organic, physical and biochemists, as confirmed by the number of citations it achieved[2] and the rapid growth in the number of publications describing investigations into alcohol–water mixtures.

In the light of developments over the past two decades, the time is ripe for a reappraisal of binary aqueous solutions of alcohols. Progress in our understanding of the structure and dynamics of aqueous solutions in general has relied on developments in four main areas:

(a) high precision microcalorimeters, densimeters and vapour pressure measuring devices have made possible the complete charting of the thermodynamics of binary mixtures, down to very low concentrations and over wide ranges of temperature and pressure;

(b) scattering methods (light, X-ray and neutron) provide information about concentration fluctuations (microheterogeneities);

(c) relaxation studies, in particular nuclear magnetic relaxation, have been successful in probing the diffusional behaviour of binary aqueous mixtures;

(d) molecular theories of liquids and solutions have been developed to a state where it is no longer necessary to treat the solvent as a continuum and to appear in the calculations solely through a macroscopic property, such as viscosity or dielectric permittivity. In particular, much effort has gone into the development of credible molecular models for liquid water which might be able to account not only for the eccentric physical properties of the substance from the ice nucleation temperature in undercooled water (233 K) to the critical temperature (647 K), but also for aqueous solutions and the role of water in chemical and biochemical processes.

The above efforts have been supported by the advent of powerful computers which has given birth to a new branch of solution chemistry: computer simulation or modelling of molecular liquids and solutions, based on assumed atomic and molecular pair interaction potentials. (A critical assessment of the presently available interaction potentials for the water dimer, as used in such simulations, can be found in the article by Finney in this volume.) It is this branch of solution physical chemistry which has grown fastest during the past decade and now boasts a productive and vociferous band of practitioners which threatens to put experimentalists and theoreticians on the defensive. Wood's thorough-going analysis of computer methods for simulating the behaviour of molecular liquids[3] has indicated the strengths of these techniques but also the many pitfalls which face the inexperienced or semi-experienced practitioner.

We thought it instructive to examine how, in the light of all this experimental and theoretical activity, our views of 'simple' binary aqueous mixtures had changed and how the questions posed in the previous review had been answered. The central role played by alcohol–water mixtures in modern technological and biochemical practice has not diminished and we hope, therefore, that this review of the progress over the past 20 years may be as timely as the previous review appears to have been.

The rapid proliferation of experimental investigations into alcohol–water mixtures and ternary (or quaternary) systems in which the alcohol–water mixture provides the solvent medium makes it imperative to limit the scope of this review to *binary* mixtures, except where a third component is used to probe the structural or dynamic properties of the binary solvent medium. As was done in the previous review, we state at the outset that this account cannot be of an exhaustive nature. We concentrate on dilute solutions where theoretical methods for the interpretation of experimental data are best developed.

2. The Intermolecular Nature of Liquid Water

Basic to any discussion of binary aqueous mixtures is an appreciation of the molecular nature of the pure components, water and alcohols. Of these, the structure of water in the liquid state has been the subject of much serious study, and even more speculation. Several of the problems which were discussed in the previous review have now been solved. Looking back to 1966, it is curious that in a section headed 'The structure of liquid water', no reference was made to the one technique, diffraction, which, of all physical techniques that have been applied to the study of water, is best able to provide a description of the time-averaged molecular structure by way of the three atom–atom radial distribution functions $g_{ij}(r)$. These functions describe the time-averaged distribution in space of the molecules in terms of the interatomic spacings O–O, O–H, and H–H, it being understood that the structure of the isolated water molecule is known in detail.

The scattering of X-rays by water is due mainly to oxygen atom pairs, although O---H interactions contribute 12% and H---H atom pairs up to 2%. On the other hand, the neutron scattering amplitudes of O and ^2H are nearly equal in magnitude, so that the neutron diffraction pattern of ^2H_2O provides information about all three atom pair distributions. This also makes the neutron radial distribution curves very complex, as discussed by Dore elsewhere in this volume.

The detailed X-ray scattering studies of Narten and Levy, extending over the temperature range 4–200 °C and summarized in reference 4, have confirmed that the short-range order which exists in liquid water is reminiscent of that in ice-I; the oxygen–oxygen pair distribution function is shown in figure 1 where it is contrasted with that of argon, the abscissa being R^*, the reduced distance of separation, where $R^* = r/\sigma$. The van der Waals (or hard sphere) radius σ has been taken as 0·282 nm for water and 0·34 nm for argon.

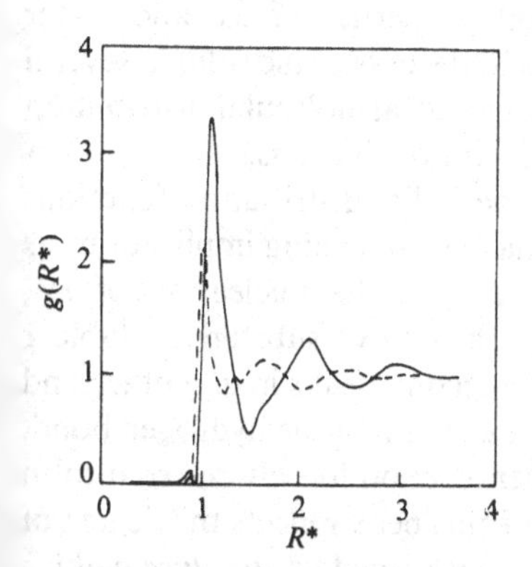

Figure 1. Radial distribution function for oxygen atoms in water (broken line) at 277 K and 100 kPa and for argon at 84.5 K and 71 kPa, as a function of the reduced distance $R^* = r/\sigma$, σ being taken as 0.282 nm for water and 0.34 nm for argon.

Table 1. *Positions of the first and second peaks in the oxygen radial distribution functions of ice and cold water*

	First peak position (nm)	Second peak position (nm)
Ice	0.275	0.450
Cold water	0.298	0.485

The fluctuations in $g(R^*)$ indicate that structure extends to about 4–5 molecular layers. The area under the first peak is a measure of the number (n_c) of nearest neighbours ($R^* \simeq 1$) of any given oxygen (or argon) atom. In the case of argon at its melting point $n_c = 10.5$ ($n_c = 12$ for hexagonally close packed spheres), decreasing with increasing temperature. For water at its melting point, $n_c = 4.4$ ($n_c = 4.0$ for ice-I), *increasing* slightly with increasing temperature. The indications are that the short-range molecular distributions in liquid water resemble those in ice and the increase in n_c with rising temperature is reflected in the phenomenon of the maximum density (4 °C for H_2O and 11 °C for ^2H_2O). Further evidence for the ice-like nature of liquid water is provided by the second peak in $g(R^*)$. For a normal fluid this is expected to centre on $R^* = 2$, as in the case of argon. For the ideal, tetrahedrally 4-coordinated structure, the next-nearest neighbour distance is given by $2\sigma \sin(1/2 \times 109° 28') \simeq 0.45$ nm; for the case of water, $R^* \simeq 1.6$. However, as argued by Finney elsewhere in this volume, the position of the second peak in $g(R^*)$ cannot be unambiguously taken as evidence of tetrahedrality, since a trigonal arrangement of oxygen atoms produces a very similar peak. What can be concluded is that the *orientations* of molecules in liquid water provide an important contribution to the detailed shape of $g(R^*)$. Nevertheless, a comparison of the physical properties of ice and water suggests that the liquid retains a resemblance to the crystalline solid in several aspects. The assumption of an approximate tetrahedral molecular distribution in the liquid is then perhaps reasonable, even if not rigorous.

An apparently minor difference between the radial distribution functions of ice and liquid water, but one which may have far-reaching implications as regards the undercooling potential of water and the nucleation of ice, concerns the lengths of the hydrogen bonds in the two substances: table 1 shows the respective positions of the first and second peaks in $g(r)$ of ice and water. The first peak is simply a measure of the length of the hydrogen bond, but the second peak is diagnostic of the phenomenon loosely referred to in the literature as *water structure*. The width of this peak reflects the degree of tetrahedrality, and ideally any solute which is described as *structure making* should produce a narrowing of this peak and an increase in the depth of the minimum between the first and the second peak. In actual fact this criterion of structure making is seldom applied in practice, mainly because the total radial distribution function of even a simple binary solution is so complex that the peak describing the distributions of next-nearest water neighbours cannot be identified. (Compared to the three radial distribution functions required for the total description of water, a solution of a monatomic solute, e.g. Ar, requires *six* such functions for a total description.)

Now that the isotopic difference technique of neutron diffraction is being applied so successfully to studies of ion hydration [**5**], there would appear to be little reason for not mounting an aqueous solution study of xenon which exists in several isotopic modifications and is reasonably soluble. In the absence of experimental data for solute–water, water–water and solute–solute distributions, we have to rely on computer simulations of such systems, and of these there is no lack. Such investigations, whether based on Monte Carlo (MC) or molecular dynamics (MD) techniques, abound in technical and computational problems. These include the details of the potential functions used, assumptions regarding pairwise additivity and ranges of atomic and molecular interactions, the size of the sample employed in the simulation and the sampling period. For instance, with one exception [**6**], the water simulations reported in the literature have been based on samples of not more than 216 water molecules. In order to study solute–solute interactions, a minimum of two solute molecules (or atoms) need to be introduced, so the system is then composed of 214 water molecules and two solute molecules, i.e. a solute concentration of approximately 0.5 mol kg^{-1}, far in excess of the measured saturation solubility of, say, argon or methane. [**7, 8, 9**]

With the advent of computer simulation methods, the consensus view regarding the best structural representation of water has shifted from mixture to continuum models in which the liquid is pictured as a macroscopically connected, random network of hydrogen bonds the topology of which is undergoing rapid and continuous transformations. The average calculated hydrogen bond distribution is shown in figure 2, indicating that two-thirds of the water molecules are bi- or trifunctional. [**10**] This gives rise to a wide variety of polyhedra with considerable hydrogen bond strain.

The comprehensive investigations, mainly by Angell and his colleagues, into the physical properties of undercooled water [**11**] have resulted in re-examinations of the available structural models. [**12, 13**] Whereas the behaviour of water over the normal liquid range has in the past been described as eccentric, the physical anomalies become increasingly marked at subzero temperatures, to the extent that most of the physical properties of the liquid appear to diverge to infinity at approximately 228 K. The nature and origin of this 'catastrophe' are still subjects of debate, but Stillinger has analysed possible structural consequences of undercooling. [**13**] He adopts as a basic structural unit the bicyclic octamer, shown in figure 3, which is the unit building block of hexagonal ice (ice-Ih). The concentration of such octamers and other unstrained polyhedra would be expected to increase at low

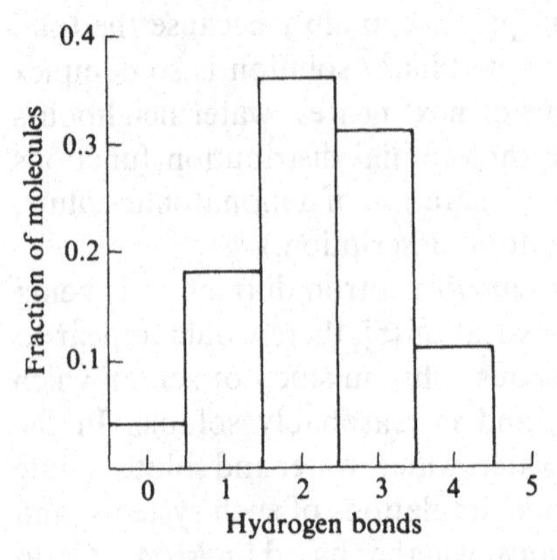

Figure 2. Fractions of molecules in liquid water at 10 °C participating in various degrees of hydrogen bonding, according to a Molecular Dynamics computer simulation.[10] The density is 1 g cm^{-3} and the critical hydrogen bond strength is taken as 17 kJ mol^{-1}. Reproduced, with permission, from reference 13.

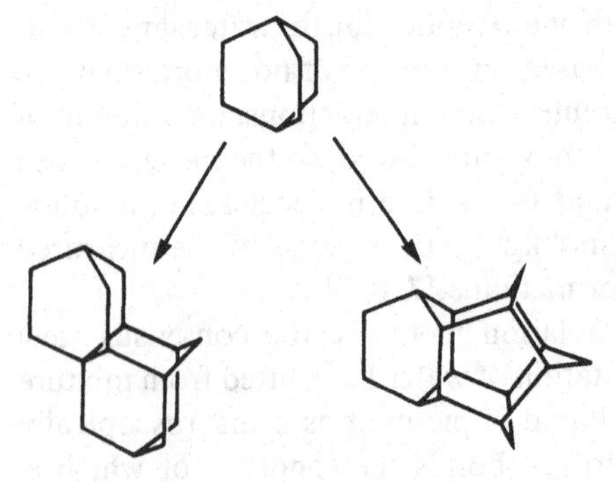

Figure 3. The bicyclic octamer unit structure, as it occurs in ice-Ih and two possible 'condensation' structures with unstrained hydrogen bonds neither of which is compatible with ice-Ih; the left hand structure does occur in ice-Ic; after Stillinger.[12]

temperatures, and since they can fuse without hydrogen bond distortion, aggregated structures of increasing dimensions are able to form. Two such possibilities are shown in figure 3: one is based on a combination of pentagons and hexagons and has no known crystalline analogue. The other one, composed of hexagons only, occurs naturally in cubic ice (ice-Ic). Yet other unstrained polyhedral structures can be conceived, based on a variety of the well-known clathrate hydrate cages.

Water is often described as ice-like, but if sufficiently large structural units composed of the bicyclic octamer unit did exist at the equilibrium freezing point, then it would be difficult to account for the ability of liquid water to undercool. We therefore assume that, whatever unstrained aggregated domains do exist at 273 K, they are either not sufficiently ice-Ih-like or they do not possess the necessary dimensions or lifetimes of a critical nucleus, i.e. one capable of spontaneous growth.

Any aggregation of well-bonded structures would clearly be of a cooperative nature and lead to large fluctuations in the density of the undercooled liquid. The average linear dimensions of low density, well-bonded domains are expressed by a correlation length (l^*) or a radius of gyration (r_g) and can in principle be obtained from small angle scattering experiments. Indeed, preliminary small angle X-ray scattering results indicate $1^* \simeq 0.8$ nm at 253 K for H_2O and somewhat larger for ^2H_2O.[14]

As the temperature is decreased towards that of the divergence catastrophe there must be an increasing probability that enough of the ice-like octamer structures can fuse to provide a critical nucleus for ice growth in the deeply undercooled liquid. The theory of homogeneous nucleation,[15] as applied to the most recent experimental data on nucleation rates, predicts that this event occurs at 233 K for droplets of micrometre dimensions at low cooling rates.[16] It would therefore seem to be very difficult, if not impossible, to probe the divergence phenomenon, unless experimental studies could be performed on rapidly quenched droplets of less than 50 nm radius.

In the meantime it suffices to say that most of the experimental results, spectroscopic, scattering, and thermodynamic alike, which have been published over the past 15 years are in general concordance with the picture of water as an infinite, irregularly hydrogen-bonded network with a distribution of bond lengths and bond strains, but in which the tetrahedral disposition of the water molecule orbitals predominates. Among the few observations which still defy an explanation are the almost pure Debye dielectric relaxation behaviour of liquid water[17] and the large (approximately 30%) difference in the self-diffusion coefficients of ^1H_2O and ^2H_2O.[18]

3. Theories of Dilute Aqueous Solutions

In recent years significant advances have been made in theoretical interpretations of experimental data for aqueous mixtures. In the past such advances had been based on theories which had originally been developed for idealized (e.g. hard sphere) fluids, but there now exist quite novel approaches particularly suitable for the study of aqueous solutions. The increasing popularity of computer simulation has provided further impetus for the theoreticians. Operationally two distinct cases must be considered, namely the infinitely dilute solution and the solution of finite concentration. The former is fully described by two pair distribution functions: $g_{11}(r)$ and $g_{12}(r)$, where species 1 is the solvent and species 2 the solute. At finite concentrations a third distribution function, $g_{22}(r)$ is required for a complete description of the system.

Historically, three major types of hydration interactions have been distinguished: (1) ion hydration where $g_{12}(r)$ is dominated by ion-dipole effects which are of a long-range nature; (2) short-range water–solute hydrogen bonding which is of particular importance in solutions of polyhydroxy compounds (e.g. sugars and sugar alcohols). Since the hydrogen bond is a very orientation-specific interaction, one must assume that the orientation dependent part of g_{12} is of importance, i.e. g_{12} (r, Ω) should be considered, and that the usual practice of orientational averaging results in the loss of much valuable information [**19**]; (3) hydrophobic hydration, which is the name given to the characteristic interaction between water and apolar atoms (rare gases) and molecules. It is believed to resemble the interactions which are responsible for the formation of clathrate hydrates and among its symptoms are some unique thermodynamic features. [**20**]

Intuitively it seems likely that both direct solute–water hydrogen bonding and hydrophobic hydration contribute to the observed properties of alcohol–water mixtures, and this is indeed found to be the case. One prime requirement for a rigorous discussion of hydration effects is a reliable pair potential $U_{12}(r, \Omega)$. Extensive *ab initio* calculations by Nakanishi and his colleagues have provided such potential functions for the pairs MeOH–water and tert.-BuOH–water. [**21, 22**] These functions subsequently served as a basis for computer simulations of infinitely dilute solutions. Only few experimental techniques are able to probe directly the concentration range which might operationally be described as infinitely dilute. Among them, thermodynamic methods have been most informative and they therefore receive pride of place in section 4.

Of more practical importance than the infinitely dilute solution is the manner in which concentration affects the solution properties. Here the theoretical approaches involve the formulation of macroscopic properties in terms of molecule–molecule potential and distribution functions. One such property is the second virial coefficient B^* in the osmotic pressure equation

$$\Pi/kT = \rho + B^*\rho^2 + C^*\rho^3 + \ldots \tag{3.1}$$

where the virial coefficients B^*, and C^* etc. measure the deviations from ideal behaviour caused by solute–solute interactions and ρ is the number density of solute molecules. B^* can be related to the solute molecule pair distribution function $g_{22}(r)$ by

$$B^* = -\frac{1}{2}\int_0^\infty [g_{22}(r) - 1]\, 4\pi r^2 \mathrm{d}r \tag{3.2}$$

where r is the distance of separation. Strictly speaking, eqn (3.2) should contain terms describing the mutual orientations of the two solute molecules. Orientation effects are expected to play a significant part in solutions of non-spherical, polar molecules and may well dominate the behaviour of molecules which are capable of hydrogen bonding. Nevertheless, for reasons of computational complexity, the orientation dependent part of g_{22} is usually neglected.

In the case of a dilute solution ($\rho \to 0$)

$$g_{22}(r) \simeq \exp[-W_{22}(r)/kT] \tag{3.3}$$

and

$$B^* \simeq -1/2 \int_0^\infty [\exp\{-W_{22}(r)/kT\} - 1] \, 4\pi r^2 \mathrm{d}r \tag{3.4}$$

where $W_{22}(r)$ is the potential of mean force, i.e. the *effective* potential between two solute molecules with the influence of the solvent molecules averaged out. $W_{22}(r)$ therefore describes solute interactions in a dilute solution (strictly speaking, so dilute that it contains only two solute molecules), just as the pair potential function $U_{22}(r)$ describes such interactions in a dilute gas. By its very definition, $W_{22}(r)$ can provide no explicit information about solute–solvent (solvation) interactions. Formally such information, e.g. $g_{12}(r)$, can only be obtained from scattering data or from thermodynamic measurements extrapolated to infinite dilution conditions. Implicitly any modification in the solvation interactions are reflected in $W_{22}(r)$ and $g_{22}(r)$, but they cannot be extracted in a rigorous manner. Many attempts have been made to calculate $W_{22}(r)$ by *ab initio* methods of greater or lesser refinement. Such data are of crucial importance in computer simulations of solutions. But, however accurate a molecular pair potential may turn out to be, there is still a large question mark against the significance of the simulation of a many-body system, e.g. two solute molecules + 214 water molecules: Can the macroscopic behaviour of such a system be adequately described by the sum of molecular pair interactions?

An examination of the various techniques currently in use for the calculation of molecular pair potentials is beyond the scope of this review, so we confine ourselves to a discussion of the results of such calculations where they relate to alcohol–water mixtures.

Friedman and his colleagues have been successful in supplanting several older chemical models of solvation by a Hamiltonian model, which has enabled them to relate solvation effects to the concentration dependence of the excess thermodynamic quantities, such as free energies, enthalpies and volumes. Initially the model was developed for ionic solutions, **[23]** but subsequently it has been used to interpret the enthalpies of mixing of alcohols in aqueous solution. **[24]**

The model interaction between two solvated solute molecules is described in figure 4. Each solute molecule is surrounded by a hydration sphere (the cosphere) in which the water molecules are distinctly different from those of water in the bulk. In other words, $g_{11}(r)$ cosphere $\neq g_{11}(r)$ bulk. The mutual approach of the solute molecules gives rise to interference between the two cospheres which can lead to the displacement of water from the cosphere(s) and/or the modification of the water remaining in the new cosphere. For simplicity's sake it is usually assumed that no such modification occurs, so

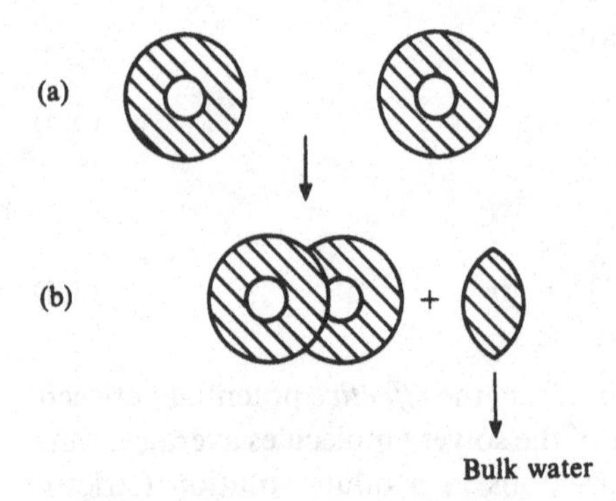

Figure 4. The cosphere overlap model of hydration: (a) hydrated molecules at infinite distance of separation, i.e. no interaction; (b) the overlap of hydration shells gives rise to solute–solute interactions and the 'extrusion' and relaxation of an equivalent volume of water to its unperturbed, bulk state. After Friedman and Krishnan.[24]

that the approach of the two solute molecules towards one another is accompanied only by a reduction in the total *number* of molecules of water in the cosphere. The volume of water so extruded can be calculated by the assignment of reasonable values to the dimensions of the hydration shell. It is now clear that the excess free energy (Helmholtz) of the solute species, contains a contribution from the cosphere overlap. The total pair potential function is written as

$$U_{22}(r) = U_{22}(r)\,\text{core} + U_{22}(r)\,\text{cosphere} \tag{3.5}$$

where the first term on the r.h.s. is a short-range core repulsion of r^{-n} type, where $9 < n < 12$. The cosphere term has a parameter, A_{22}, which is adjusted to produce a fit to the experimental ΔA_e data. Most simply expressed, A_{22} is the free energy change per mol of water displaced when the cospheres of the solute molecules are brought to a state of overlap from $r = \infty$. One might therefore also associate A_{22} with changes in the hydration free energy of the solute species and thus indirectly obtain information about hydration from thermodynamic excess functions, such as ΔA_e and its temperature and pressure derivatives.

The model, as outlined so far, has been successful in describing aqueous solutions of monatomic ions, but problems arise with polar molecules, especially those, like alcohols, which can participate in hydrogen bonding both as proton donors and acceptors. It is conceivable that the cosphere associated with the –OH group does not resemble that of the alkyl group or that the cospheres of isomers such as 1-BuOH and tert.-BuOH could differ significantly, so that in mixtures of species 2 and 3 $A_{22} \neq A_{33} \neq A_{23}$. Despite such shortcomings, the cosphere overlap model has made significant contributions to our current understanding of dilute aqueous solutions.

Just as ΔA_e is related to $W_{22}(r)$ and $g_{22}(r)$, so its pressure and temperature derivatives ΔS_e and ΔV_e are directly related to the corresponding derivatives of $W_{22}(r)$ and $g_{22}(r)$. By means of the cosphere overlap model the various derivatives of A_{22} can be calculated from the thermodynamic excess functions.

The Kirkwood–Buff theory[25] provides another means for relating structure, as revealed by scattering, to the thermodynamic properties of a mixture. The theory, first developed in 1951, never became very popular with solution chemists until fairly recently.[26] We present here its basic features. In a binary mixture the average excess of the number of 2 molecules to be found in a spherical shell around a 1 molecule over the statistical average number of 2 molecules is

$$\rho_2[g_{12}(r)-1]\,4\pi r^2\mathrm{d}r. \tag{3.6}$$

The space integral is given by

$$\rho_2 G_{12} = \rho_2 \int_0^\infty [g_{12}(r)-1]\,4\pi r^2\mathrm{d}r \tag{3.7}$$

and G_{12}, the so-called Kirkwood–Buff integral, is the average excess over the bulk average of the number of molecules of type 2 (or 1) around a molecule of type 1 (or 2) per unit number density of molecules of type 2 (or 1). (The quantity G_{ij} must not be confused with the Gibbs free energy function, also commonly denoted by G.)

For a one-component liquid G can be expressed completely in terms of the thermodynamic properties:

$$G = kT\beta_T - \rho^{-1} \tag{3.8}$$

where β_T is the isothermal compressibility coefficient $-V^{-1}(\partial V/\partial P)_T$. For a binary mixture the various G_{ij} can be expressed by the composition dependence of the thermodynamic quantities, i.e. by the partial molar quantities. For a mixture of n_i and n_j mols of the two components, respectively,

$$G_{ii} = kT\left(\beta_T + \frac{\bar{V}_j^2}{V\mu_{jj}}\right) - \rho_i^{-1} \tag{3.9}$$

and

$$G_{ij} = kT\left(\beta_T + \frac{\bar{V}_i \bar{V}_j}{V\mu_{ij}}\right) \tag{3.10}$$

where V is the molar volume of the mixture, $\bar{V}_i$ and $\bar{V}_j$ the partial molar volumes of the components, $\rho_i = n_i/V$, and $\mu_{ij} = (\partial\mu_i/\partial n_j)$, μ_i and μ_j being the respective chemical potentials.

The condition for an ideal solution is

$$\Delta = G_{11} + G_{22} - 2G_{12} = 0$$

for all compositions.

Now the intensity of scattered radiation (light, X-rays or neutrons), when extrapolated to zero scattering angle, is related to the linear combination of the G_{ij} integrals. Thus, if scattering data were available over a large range of angles then the G_{ij} values obtained from thermodynamic measurements could be related to $g_{ij}(r)$, as obtained from wide-angle scattering measurements.

Donkersloot has examined the relationship between $g(r)$ and the Kirkwood-Buff integral G, given by eqn (3.7), for liquid water and has extended the treatment to various binary mixtures [**27**]; his results, as they relate to aqueous solutions will be reviewed in section 6.

The nature of the G_{ij} integrals as functions of composition, molecular dimensions and interaction energies in mixtures of Lennard–Jones 6–12 fluids has been calculated with the aid of the Percus–Yevick theory. [**28**] For such fluids

$$U_{ij}(r) = 4\epsilon_{ij}[(\sigma_{ij}/r)^{12} - (\sigma_{ij}/r)^{6}] \tag{3.11}$$

and it is often assumed that

$$\sigma_{12} = (\sigma_{11}+\sigma_{22})/2 \quad \text{and} \quad \epsilon_{12} = (\sigma_{11}\sigma_{22})^{1/2}. \tag{3.11a}$$

In mixtures where the intercomponent dimensions and interaction energies can be so expressed, neither G_{11} nor G_{22} exhibit maxima at any composition. However, when $U_{12} > U_{11}$ or U_{22}, maxima develop in G_{11} and G_{22} and a minimum in G_{12}. The experimental data for mixtures of cyclohexane + 2,3-dimethylbutane can be fitted very well by optimizing the various parameters in eqn (3.11) and the packing fraction, defined by the Percus–Yevick theory as

$$\eta = \sum_{i} (\Pi/6)\rho_i \sigma_{ii}^3. \tag{3.12}$$

For mixtures of polar and non-polar molecules (acetonitrile + toluene) the experimental G_{ij} curves are complex. For instance, G_{11} and G_{22} exhibit shallow minima (not at the same composition), while G_{12} has a marked minimum at $x_{\mathrm{MeCN}} \simeq 0.7$ (where x is the mol fraction). Optimization of all the σ_{ij} and η parameters cannot produce an adequate fit to the data, although most of the qualitative features in G_{ij} can be reproduced.

As U_{11} and U_{22} increase relative to U_{12}, and the system tends towards phase separation, so G_{11} and G_{22} increase steeply and G_{12} decreases correspondingly. This type of behaviour is observed for alcohol–water mixtures and is confirmed by scattering results (see below). For obvious reasons such mixtures cannot be modelled by hard-sphere or Lennard–Jones fluids except by the assignment of quite unrealistic values to the quantities in eqns (3.11) and (3.12). It is interesting, nevertheless, that non-monotonic $G_{ij}(x)$ behaviour can apparently be expected even for fairly simple mixtures which do not deviate greatly from the conditions in eqn (3.11a).

Finally, and for the sake of completeness, we briefly review chemical models which have from time to time been applied to explain the physical properties of complex aqueous mixtures. Thus Kozak, Knight and Kauzmann in their classic paper[**29**] tested the applicability of the lattice model, originally developed by Flory to describe the behaviour of polymers in solution[**30**], and adapted by Guggenheim for mixtures of small molecules[**31**]. The objective is to predict the virial coefficients in the activity equation:

$$\ln a_w = Bx^2 + Cx^3 + \ldots \tag{3.13}$$

where B, C etc. are related to B^*, C^* etc. in eqn (3.1) and account for non-ideality. Formally B and C describe interactions between pairs and triplets of solute molecules.

The simple lattice model, based on athermal mixing, predicts that $B < 0$ and $C > 0$, irrespective of the assumed dimensions of the solute molecules, as expressed by the number of lattice sites which they occupy.

The model can be refined by including hydration effects in the calculations. A notional hydration number n_h is assigned to the solute molecule, representing the number of water molecules associated with a solute molecule. The effect of such hydration on a_w can then be calculated for very dilute solutions.[**32**] It invariably leads to B and $C < 0$. Finally, if the restriction of athermal mixing is lifted and molecular pair interaction energies are introduced, it can be shown that $B \gtrless 0$ but $C < 0$.

The experimental results are at odds with the predictions of the lattice model, since both positive values of B and C are observed; similarly dB/dT and dC/dT can assume both positive or negative values.[**29**] Patterns can be detected for members of a given homologous series but not between different series. Thus, for the alkanols at 0 °C, $B > 0$ and $C < 0$ (except for MeOH), $dB/dT > 0$ and $dC/dT > 0$.

More refined chemical models have from time to time been proposed to rationalize the observed physical properties of alcohol–water mixtures. Each is based on a number of postulated equilibria which govern the co-existence of various distinguishable species in solution. Such models are not likely to lead to a real understanding of the molecular interactions in aqueous mixtures. They are nevertheless valuable because they allow predictions to be made of the physical properties of complex aqueous mixtures over the whole composition range. Some of these chemical models and their applications are reviewed in section 6. At this stage we emphasize that they are exercises at fitting experimental data with a minimum number of adjustable parameters and they are not to be compared with *ab initio* models.

4. Experimental Studies

It is impossible, within the compass of this review, to provide a comprehensive catalogue of the experimental investigations performed on alcohol–water mixtures over the past two decades, nor is it necessary. We shall focus on some representative studies in the major areas of physical chemistry; these include structure, energetics and dynamics.

4.1. *Scattering of Radiation*

The most direct approach to structure is by scattering experiments which yield information about distances and angles, the very elements of structure. The interpretational complexities are such that no wide-angle X-ray or neutron scattering results are yet available for alcohol–water mixtures. The only indication of structure in such mixtures comes from an examination of the concentration dependence on the intensity of the X-ray diffraction band observed for aqueous solutions. [33] For mixtures of ethanol and water the scattering intensity for all values of x lies above the calculated intensity, based on the sum of the intensities of the two pure components. Furthermore, the intensity exhibits a sharp peak at $x = 0.1$ and a broader peak near $x = 0.75$.

The structural information provided by light scattering and small angle X-ray (SAXRS) and neutron (SANS) scattering experiments is not as detailed as that which could, in principle, be derived from wide angle scattering, but the data processing is more direct and the techniques have been used to probe elements of structure in aqueous solutions.

Fujiyama *et al.* have performed detailed light scattering studies on tert.-BuOH–water mixtures. [**34**, **35**, **36**] Their findings highlight several general features of alcohol–water mixtures, so that we shall discuss the results in some detail. In the context of this review, the chief value of light scattering data lies in the relationship between the Rayleigh scattering intensity and the mean square concentration fluctuations $N(\Delta x)^2$ which give rise to the scattering. Thus,

$$N(\Delta x)^2 = \frac{N\rho\mathcal{N}\lambda_i^4}{\overline{M}}\frac{1}{\pi^2}\frac{1}{4n^2(\partial n/\partial x)^2_{T,P}}\{(R_{90})_R-(R_{90})_S\} \tag{4.1}$$

where N is the number of molecules in the volume within which the concentration fluctuation is measured, $\mathcal{N}$ is the Avogadro number, λ_i the incident wavelength of the light, n the refractive index, ρ the density and $\overline{M}$ the mean molecular weight of the mixture. $(R_{90})_R$ is the measured Rayleigh scattering intensity relative to that of pure chloroform and $(R_{90})_S$ is the corresponding ratio due to entropy fluctuations. The latter can be calculated and depends only on certain physical properties of the mixture and on the temperature. [**34**] The concentration fluctuations, as defined by eqn (4.1) and obtained from the experimental $(R_{90})_R$ data, are shown in figure 5 where they are compared to similar data for 1,4-dioxan–water mixtures and to the expected values for an ideal binary mixture.

The following features emphasize the remarkable behaviour of alcohol–water mixtures: (i) the mean square fluctuations (F) increase dramatically

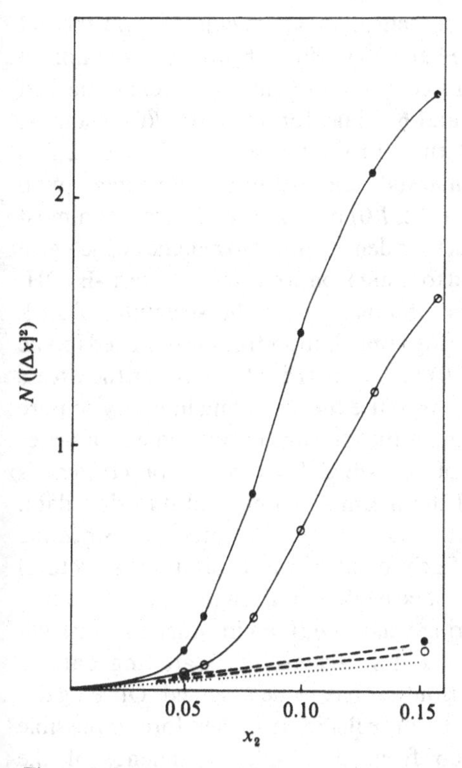

Figure 5. Concentration and temperature dependence of the mean square concentration fluctuations in tert.-BuOH–water (solid line) and dioxan–water (broken line) mixtures at 17.5 °C (open dot) and 49.5 °C (solid dot). The dotted line indicates the expected fluctuations for an ideal solution. Data from reference 36.

beyond the region $x \leqslant 0.05$. Let x_c be the actual concentration at which $dF(x)/dx$ shows the sharp rise, then $(dx_c/dT) < 0$; (ii) for $x < x_c$, $F(x)$ is temperature-insensitive but for $x > x_c$, $dF(x)/dT \gg 0$, a behaviour characteristic of lower critical demixing. By contrast the $F(x)$ curves for dioxan–water mixtures do not begin to exhibit significant deviations from ideality until $x > 0.15$ and they are not markedly affected by temperature. We defer the interpretation of figure 5 to section 6 but conclude at this stage that dilute aqueous solutions of tert.-BuOH exhibit symptoms of lower critical demixing, although no macroscopic phase separation does actually take place in such mixtures.

Two SAXRS investigations have been performed on tert.-BuOH–water systems. Bale *et al.* measured the angle and composition dependence of the scattering intensities for $0 < x < 0.4$ at three temperatures.[37] Atkinson, Clark and Franks (1975, unpublished results) performed similar measurements with greater emphasis on the dilute solution range. The zero angle scattering intensities, $I(0)$, were obtained by standard extrapolation methods and found to agree reasonably well with the theoretical values calculated with the aid of the Kirkwood–Buff integrals; see eqn (3.7). $I(0)$ is shown as a function of x at three different temperatures in figure 6. As might be expected, marked similarities exist between figures 5 and 6. Thus, for $x < 0.05$, $I(0)$ is almost constant and insensitive to temperature. In the range $0.05 < x < 0.12$ $I(0)$ increases rapidly and shows a marked temperature dependence, with $dI(0)/dT > 0$. After peaking at $x \approx 0.12$, $I(0)$ decreases and becomes almost independent of temperature. The electron density inhomogeneities which give rise to the observed scattering are said to result from a clustering of tert.-BuOH, but no detailed information can be obtained about the structure of such clusters. In the first place, the scattering contribution from water needs to be subtracted from the experimental $I(0)$ values in order to arrive at the excess electron density. A common practice is to use the scattering intensity of pure water for this purpose, but this assumes that the local arrangement of water molecules in the neighbourhood of tert.-BuOH molecules or clusters is identical to that in pure water. All the thermodynamic and diffusion data, still to be discussed, suggest such an assumption to be untenable. Since the *excess* scattering by the mixtures is fairly small, any changes in the structural correlations of water molecules cannot be neglected. Even if the usual method of correcting for the solvent scattering was valid, then in order to construct models of the tert.-BuOH clusters to test against the scattering data, a uniform cluster size and shape distribution has to be assumed. Once again, such a procedure would appear to be unrealistic. It is therefore impossible to extract quantitative information from the angle dependence of the

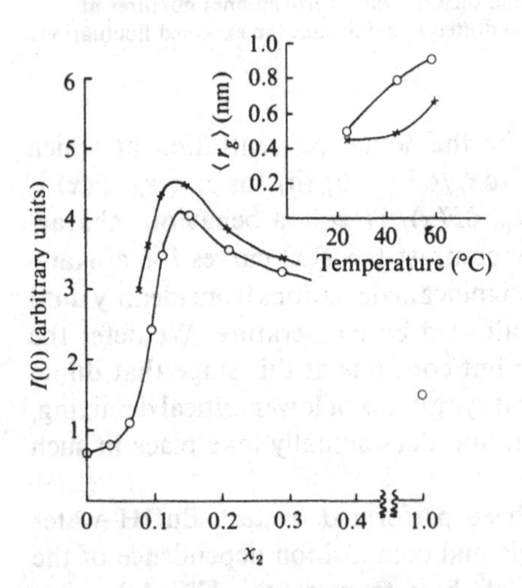

Figure 6. Zero angle X-ray scattering intensities $I(0)$ of tert.-BuOH–water mixtures at 25 °C (open circle) and 60 °C (cross), according to Atkinson and Franks (unpublished results, see text). Inset: mean radius of gyration as function of temperature for mixtures in which $x_2 = 0.1$ (open circle) and 0.21 (cross).

scattering curves, but their general appearances undoubtedly indicate that the concentration fluctuations are confined to the composition range $0.05 < x < 0.4$, peaking at $x \approx 0.12$.

An approximate estimate of the cluster dimensions can be obtained from the angle at which the scattering curve becomes angle dependent. This is expressed as $\langle r_g \rangle$, the radius of gyration (or a correlation length) in the inset to figure 6. The dimensions of the clusters are seen to approach those of surfactant micelles, indicating extensive microheterogeneity in the system. The amplitude of the density fluctuations in pure water have already been referred to in section 2 and it was seen that $\langle r_g \rangle$ *increases* with *decreasing* temperature to approximately 0.4 nm at 253 K in undercooled water. Bosio *et al.* have reported SAXRS measurements on undercooled aqueous solutions of ethanol.[**14**] They found that $\langle r_g \rangle$ becomes 'dramatically smaller' when, what they call a 'hydrogen-bonding impurity' is present, the reduction in $\langle r_g \rangle$ increasing with the ethanol concentration. Thus it appears that ethanol at low concentrations ($x < 0.06$) reduces the dimensions of the low density domains in water. Ethanol then acts in the same sense as a rise in temperature.

At first sight this contradicts the light scattering and SAXRS data on tert.-BuOH–water mixtures. However, results from thermodynamic measurements clearly indicate that for ethanol $x_c \approx 0.08$ (0.04 for tert.-BuOH), and no alcohol clustering is observed for $x < x_c$. The findings of Bosio *et al.* do not therefore refer to the same type of system as do the results on the tert.-BuOH–water systems discussed above.

The combination of the available scattering data suggests that for $x < x_c$ the main effect of the alcohol is to interfere destructively with the low density domains in water, whereas for $x > x_c$ the predominant effect becomes one of compositional micro-heterogeneities, leading to alcohol clustering.

4.2. *Thermodynamic Properties*

Thermodynamic properties are the starting point for many investigations of solutions or liquid mixtures. They can be defined rigorously and measured quite precisely. They permit the equilibrium state (e.g. solubility, vapour pressure) to be calculated as a function of temperature or pressure. They are macroscopic properties in that they only give information on the states of the system as a whole averaged over space, time and species. They are, however, ideal for quantitative tests of the various solution models and theories that have been proposed.

Extensive data exist on the thermodynamic properties of alcohol–water mixtures. The earlier studies primarily covered high concentrations and emphasized the thermodynamic mixing functions. In the last 20 years, however, major efforts have been made to extend these investigations to dilute solutions where the most striking changes in properties are observed. No attempt will be made here to review the total literature in this area. We will rather concentrate on major trends in the properties with concentration, chain length, temperature and pressure, with particular emphasis on dilute solutions.

4.2.1. *Thermodynamic Relations for Binary Systems.* It is customary to study liquid mixtures by means of mixing functions, defined by

$$\Delta Y_{\mathrm{m}} = (Y - n_1 Y_1^* - n_2 Y_2^*)/(n_1 + n_2) \tag{4.2}$$

where Y is the total measured property, Y_1^* and Y_2^* are the molar properties of the pure water (1) and alcohol (2) and n_1 and n_2 are the respective number of mols of both components. These mixing functions are the excess functions for all properties except free energies and entropies. In these two latter cases the *ideal* functions of mixing must be subtracted from ΔY_{m} to obtain the excess functions, i.e. the non-ideal contributions.

This procedure well describes the overall non-ideality of a system with respect to Raoult's law. However, this is by no means the best way to describe all the various interactions in the system. The mixing functions use as reference states the two pure liquids, but the infinite dilution states are often more informative for the investigation of solute–solvent and solute–solute interactions. This is particularly important with alcohols since the properties of their aqueous solutions exhibit very pronounced variations in the water-rich region. In such instances partial molar quantities are the preferred functions. For a binary system, they are defined by

$$Y_1 = (\partial Y/\partial n_1)_{n_2, T, P}; \quad Y_2 = (\partial Y/\partial n_2)_{n_1, T, P}. \tag{4.3}$$

Y_1 and Y_2 are formally related by the Gibbs–Duhem equation. Y_2 becomes very sensitive to the various interactions as $x_2 \rightarrow 0$. On the other hand, Y_1 is very insensitive under these conditions, because when $x_1 \rightarrow 1$, then $Y_1 \rightarrow Y_1^*$.

Physically, the interactions in the system are distributed over few molecules near infinite dilution and over a large number of molecules near the pure liquid state. Ideally, the maximum information in a binary system will be obtained if Y_1 and Y_2 are examined for both solutes at low concentration and ΔY_{m} in the intermediate zone.

With few exceptions, Y_i cannot be measured directly. It is usual to calculate Y_i from apparent molar quantities $\phi_{Y,i}$ or from ΔY_{m}. By definition, for solute 2,

$$\phi_{Y,2} = (Y - n_1 Y_1^*)/n_2. \tag{4.4}$$

It follows from eqns (4.3) and (4.4) that

$$Y_2 = \phi_{Y,2} + n_2(\partial \phi_{Y,2}/\partial n_2)_{n_1, T, P}. \tag{4.5}$$

In dilute solutions, n_2 can be replaced by the molality m. For more concentrated solutions mol fractions are more convenient and

$$Y_2 = \phi_{Y,2}^0 + x_1 x_2(\partial \phi_{Y,2}/\partial x_2)_{T, P}. \tag{4.6}$$

If $\phi_{Y,2}$ is written as a polynomial in m or x_2, then Y_2 is obtained directly from eqn (4.5) or (4.6). It can also be obtained graphically from a plot of $\Delta(\phi_{y,2}m)/\Delta m$ against the mean molality or from a plot of $\Delta(\phi_{y,2}x_2/x_1)/\Delta(x_2/x_1)$ against the mean mol fraction. Similar relations can readily be written for the other component, water.

If the available data are in the form of ΔY_m, then

$$\phi_{y,2} = (\Delta Y_m/x_2) + Y_2^* \tag{4.7}$$

or

$$Y_2 = \Delta Y_m + x_1(\partial \Delta Y_m/\partial x_2) + Y_2^*. \tag{4.8}$$

The standard partial molar quantities for a solute $Y_2{}^0$ are, with the exceptions of free energies and entropies, the values at infinite dilution. These standard values depend only on the intrinsic contribution of the molecule itself and on solute–solvent interactions. It is often convenient, and sometimes essential (e.g. enthalpies), to eliminate the intrinsic contribution. This can be done through hydration functions:

$$\Delta Y_h^0 = Y_2^0 - Y_2^0 \text{ (gas phase)}. \tag{4.9}$$

When it is not possible to obtain gas phase data, e.g. for the volumetric properties, the difference $Y_2^0 - Y_2^*$ is often convenient for a discussion of solute–solvent interactions.

The adoption of the pure liquid alcohol as reference state, although convenient, may give rise to problems when the properties of different alcohols, each at infinite dilution, are compared at the same temperature, usually 298.16 K. Thus, methanol is close to its boiling point, while tert.-BuOH is only just above its melting point (26 °C). The molecular states and interactions in the two pure liquids are likely to be very different, and this will be reflected in $(Y_2^0 - Y_2^*)$, making such comparisons of questionable value. Franks *et al.* have suggested that comparisons of thermodynamic properties at a common *reduced* temperature might be more appropriate [**38**], but there are of course limitations imposed by the temperature range (253–373 K) over which a liquid mixture can exist at ordinary pressures.

At low concentrations, Y_2 can often be expressed as a virial expansion in m:

$$Y_2 = Y_2^0 + y_{22}m^2 + y_{222}m^3. \tag{4.10}$$

The parameters y_{22}, y_{222}, etc. are second, third and higher virial coefficients and are related to B, C etc. in eqn (3.13) when $Y = G$ (the Gibbs free energy). In the case of free energies and entropies, extra terms ($RT \ln m$ for G_2 and $-R \ln m$ for S_2) must be added to account for the ideal functions of mixing. The parameter y_{22} is of great interest since it is related to pair interactions between two solute molecules as they move from an infinite distance to some critical short distance of separation which is not necessarily equal to the van der Waals diameter of an alcohol molecule. The parameter y_{22} will be the averaged sum of all solute–solute interactions and is related to the integral in eqn (3.4) for the case where $y_2 = \mu_2$, the chemical potential. Modifications of the solute–solvent interactions as two solutes approach each other is also implicitly considered as a solute–solute interaction, see, for instance eqn (3.5).

The following sections summarize recent thermal and volumetric investigations into alcohol–water mixtures. With very few exceptions such studies have been confined to the lowest members of the homologous series (C_1 to C_4), with emphasis on EtOH and tert.-BuOH. Corresponding data for the higher homologues are frequently calculated from a constant CH_2 contribution to the relevant quantity, as obtained from one of the several group additivity schemes which have been proposed, see for example Cabani *et al.*[39] The accuracy of such estimated data is often comparable with the experimental uncertainty.

4.2.2. *Free Energies, Enthalpies, and Entropies.* Excess free energies are usually derived from measurements of colligative properties, while excess enthalpies are obtained from calorimetric heats of mixing. From the free energies and enthalpies, excess entropies are readily obtained. Most of the early results have been adequately reviewed.[1] The more recent data have been analysed by Westmeier[40] who also tabulated the most reliable values for free energies, enthalpies, volumes and viscosities of the alcohols that are completely miscible with water at 25 °C. In principle it is sufficient to measure the free energies and enthalpies at only one temperature, since it is then possible to calculate their temperature dependence from measured heat capacities and the pressure derivative of the free energies from measured volumetric properties. Even so, the low temperature activity data of Mitsutake and Sakai[41] and the high temperature activity data of Pemberton and Mash[42] of aqueous ethanol solutions are worth noting. High temperature and pressure liquid–vapour equilibria in methanol–water systems have been investigated by Pryanikova and Efremova[43], while Larkin[44] has reported the temperature dependence of the excess enthalpies of water–ethanol mixtures.

Very precise low concentration activities of aqueous alcohol solutions have recently been measured with the aid of a novel freezing point technique[45] and have yielded the first accurate second virial coefficients [eqn (4.10)]. Christian and Tucker[46] performed extremely precise vapour pressure measurements on aqueous methanol solutions down to very low concentrations, for the same purpose.

The aim generally is to extend the measurements to concentrations that are low enough for higher terms in eqn (4.10) to contribute only insignificantly to Y_2. The actual concentration range where this is the case depends primarily on the alkyl chain length of the alcohol. Needless to say, the above requirement makes rigorous demands on the precision of the chosen experimental technique.

Enthalpies of solution and of dilution of alcohols in dilute aqueous solution have been measured by Friedman and Krishnan [**24**], Franks *et al.* [**47**], Desnoyers *et al.* [**48**] and Perron and Desnoyers. [**49**] Such studies have been facilitated by the availability of various types of highly sensitive microcalorimeters. The results, when combined with enthalpies of mixing, allow the calculation of partial molar relative enthalpies of alcohol and water over the whole mol fraction range. Franks *et al.* [**47**] also measured the enthalpies of mixtures of two alcohols in water at low total concentrations and hence derived enthalpy pair interaction parameters between different alcohols.

4.2.3. *Volumes.* Volumetric data are generally obtained from density measurements, and with modern techniques densities can be measured readily to a few parts per million (p.p.m.). Partial molar volumes can thus be obtained with reasonable precision for solutions as dilute as 10^{-2} mol kg^{-1}. Most of the information about expansibilities has been deduced from the temperature dependence of densities, but dilatometry can also be used effectively for this purpose. A recently developed flow dilatometer [**50, 51**] is interesting in this respect.

Density data on alcohol–water mixtures of sufficient precision for the calculation of ΔV_m have been available for some time and Westmeier [**40**] has reviewed the density literature, but limited to data at 25 °C. Density determinations over the whole composition range but also precise enough for the derivation of partial molar quantities at low concentrations are of more recent origin. [**52–60**] Less precise data, but covering a large domain of temperature and pressure have been reported for methanol and ethanol. [**62, 63**]

The importance of accurate volumetric data for dilute aqueous solutions of alcohols has been stressed by Franks *et al.* [**64, 65**] The chief problem concerns the reliable extrapolation of $V_2(x_2)$ to its limiting value V_2^0. For electrolytes the Debye–Hückel limiting law prescribes the correct extrapolation procedure, but in the absence of a similar law for non-electrolytes the extrapolation must be performed by experiment, i.e. a concentration range must be established in which $(\partial V_2/\partial x_2)$ becomes linear or possibly equal to zero. [**65**]

In recent years, volumetric measurements on alcohol–water mixtures (and aqueous solutions of other non-electrolytes) have concentrated on the estimation of V_2^0 [**66–70**], the virial coefficients in eqn (4.10) [**71**], the detailed shape of $V_2(x_2)$ as a function of alkyl chain length [**72**] and the feasibility of expressing V_2^0 and V_2 as sums of functional group contributions. [**39, 72, 73**]

The volumetric properties of dilute ternary aqueous solutions of alcohols have also been studied [**49, 74**] and so have the volumetric properties of water in its dilute solution in alcohols. [**75, 76**]

4.2.4. *Heat Capacities, Expansibilities and Compressibilities.* We have so far considered measurements of the first temperature and pressure derivative of the Gibbs free energy, namely H and V. We now turn our attention to the second T and/or P derivatives. Contrary to other thermodynamic functions, relatively few precise studies of heat capacities and compressibilities were available until refined calorimetry and sound velocity techniques were developed in the early 1970s. The properties that are usually measured experimentally are isobaric heat capacities and isentropic (adiabatic) compressibilities. From these data heat capacities at constant volume and isothermal compressibilities can be calculated, provided that reliable expansibility data are available. Unfortunately, these have often been lacking in the past.

Precise apparent and partial molar isentropic compressibilities, usually at 298 K, and extending over the whole miscibility region are now available [**77–79**]; compressibilities of tert.-BuOH–water mixtures have also been investigated as a function of pressure. [**80**]

The dilute aqueous region has been investigated with special care, the aim being to obtain details of alcohol hydration and the derivation of group additivity schemes for isentropic and isothermal standard partial molar compressibilities. [**66, 81–84**] In many cases, temperature effects were also investigated.

Precise heat capacities over the whole mol fraction range are available, often at several temperatures, for most simple alcohol–water mixtures. [**55, 56, 61, 85–88**]. The low concentration range has also been thoroughly investigated. [**73, 89, 90**] Group additivity schemes have been suggested by several workers. [**39, 73, 90–92**] Specific heats of aqueous ethanol solutions have been determined over a wide temperature range, see for example Kessel'man *et al.* [**93**] and Shal'gin and Puchkov [**94**], although such measurements are not precise enough for the derivation of partial molar quantities of dilute solutions. Heat capacity data on mixtures of two alcohols in water have been reported by Perron and Desnoyers [**49**] and heat capacities of water in alcohols have been measured by DeGrandpré and Jolicoeur (unpublished results).

From a theoretical point of view, C_V and K_T are more important than C_P and K_S but they are much more difficult to measure directly with an acceptable degree of precision. They are related via the coefficient of thermal expansion (α_0), the isothermal compressibility (β_0) and the isobaric heat capacity (σ_0) per unit volume of pure water, as follows:

$$K^0_{T,2} = K^0_{S,2} - \frac{2\alpha_0 T}{\sigma_0} E^0_2 - \frac{\alpha_0^2 T}{\sigma_0^2} C^0_{P,2} \tag{4.11}$$

$$C^0_{V,2} = C^0_{P,2} - \frac{2\alpha_0 T}{\beta_0} E^0_2 - \frac{\alpha_0^2 T}{\beta_0^2} K^0_{T,2}. \tag{4.12}$$

(Note: the units are balanced through the factor 10^5 cm^3 Pa = 0.1 J.)

As mentioned above, the paucity of reliable expansibility data has limited the number of calculations of isothermal compressibilities (K_T) and isochoric heat capacities. Still, it has been shown that $K_{T,2}$ is not remarkably different from $K_{S,2}$[**66, 78, 84, 85**], at least in the normal temperature range. Except for the work of Kiyohara and Benson[**77**] few high precision isochoric heat capacity measurements have been reported. Nevertheless, the trends with concentration appear to be similar for $C_{V,2}$ and $C_{P,2}$ and at infinite dilution their magnitudes are similar. Once again, this statement holds true for the normal temperature range for such measurements (280–310 K). From eqns (4.11) and (4.12) it follows that at the temperature of maximum density, where $\alpha = 0$, $K^0_{S,2} = K^0_{T,2}$ and $C^0_{P,2} = C^0_{V,2}$. For undercooled water, C_P and C_V diverge sharply with decreasing temperature [**11**] and this must be reflected in $K^0_{T,2}$ and $C^0_{V,2}$. The temperature range corresponding to the undercooled solutions remains to be explored, but the density measurements performed by Sorensen on aqueous solutions of EtOH, tert.-BuOH and hydrazine down to 253 K (and below in some cases) are worthy of note.[**95**] Although the experimental precision was by no means adequate for the calculation of reliable E_2 or E^0_2 values, the results have nevertheless thrown into sharp relief the intermolecular complexity of undercooled water and aqueous solutions.

4.2.5. *Solubilities and Phase Diagrams.* While the phase diagrams of alcohol–water mixtures have been available for many years, Moriyoshi *et al.*[**96–99**] have extended some of these diagrams to high pressures and temperatures. Ott *et al.* [**100**] and Anisimov *et al.* [**86**] have redetermined the low temperature solid/liquid phase diagrams of many alcohol–water mixtures.

4.2.6. *Other Thermodynamic Properties.* Amongst the other thermodynamic properties which can give useful information on alcohol–water interactions, interesting studies have been made on the effect of alcohols and other selected solutes on the temperature of maximum density (θ) of water[**100–103**]. $\Delta\theta$ is related to $\phi_{E,2}$, the apparent molar expansibility of the solute by the expression[**104**]

$$\alpha^0_2 = \frac{1}{V^0_2}\frac{\partial V^0_2}{\partial T} = \frac{-2.81 \times 10^{-4}}{V^0_2}\left(\frac{\Delta\theta}{x_2}\right)_{x_2\to 0}. \quad (4.13)$$

The constant 2.81×10^{-4} allows for the volumetric behaviour of water in the neighbourhood of θ, based on the assumption that $V^*_1(T)$ can be expressed by a parabola. The measurement of $\theta(x_2)$ offers an alternative and convenient method for the estimation of E_2 data of reasonable precision. Similar analyses of the temperature of minimum K_S have also been reported.[**105**]

As mentioned earlier, the temperature of maximum density θ of the solution is related to the E^0_2 of the alcohol. $\Delta\theta$ can be separated into ideal and non-ideal contributions[**101, 102, 104**] and the non-ideal contribution seems to give an indirect measure of the overall effect of the solute on the intermolecular order of water.

From the *PVT* data, it is possible to derive the energy–volume coefficients by the relation

$$(\partial U/\partial V)_T = T(\partial P/\partial T)_V - P. \tag{4.14}$$

MacDonald *et al.* have measured these coefficients for some alcohols in water and derived from them cohesive energy densities and related thermodynamic quantities. [**106, 107**]

4.2.7. *Trends in Thermodynamic Properties.* The principal quantities and trends which characterize the thermodynamic behaviour of alcohol–water mixtures can be separated into three groups. The standard thermodynamic functions which usually correspond to infinite dilution depend only on solute–solvent interactions. The second virial coefficients, obtained from the initial slope of the thermodynamic properties, measure the molecular pair interactions between solutes. For higher order interactions and observations of structural changes in the mixture, the concentration dependent trends of the partial or apparent molar quantities are often more informative than the excess mixing functions.

The main standard functions for the alcohols are summarized in table 2. Unless otherwise stated, these data were taken from the references in the preceding sections. In many cases slightly different values have been reported by different authors and the quantity given is either an average value or the value which seems more consistent with the other alcohols. The uncertainty is especially large with expansibilities since V_2^0 does not vary significantly with temperature and the E_2 values obtained indirectly are often of the same magnitude as the experimental uncertainty. Direct measurements would certainly be useful, especially in view of the importance of this function in calculating $K_{T,2}^0$ from $K_{S,2}^0$ and $C_{V,2}^0$ from $C_{P,2}^0$.

The corresponding functions for the pure liquid alcohols are included in most cases for comparison. These were taken from the same sources as the data for the aqueous solutions or from the compilation by Wilhoit and Zwolinski. [**108**] The hydration functions, ΔG_h^0, ΔH_h^0, and ΔS_h^0 (transfer from the gas phase to the aqueous solution in the standard states) were taken from the compilation by Cabani *et al.* [**39**]

The standard thermodynamic functions of the alcohols given in table 2 are all significantly different from the corresponding values of the pure liquid alcohols or of alcohols in non-aqueous solvents. Therefore, any model for solute hydration effects will have to account for all these observations. The sum of the hydration and the evaporation functions correspond to the changes in free energy, enthalpy and entropy of the alcohol when transferred from the pure liquid state to the aqueous medium. These changes are accompanied by nonideal decreases in enthalpy and entropy, compatible with the postulated increase in the structural order of water. [**64**] The heat capacities $C_{P,2}^0$ and $C_{V,2}^0$ are large and positive, indicating that such extra structure is destroyed as the temperature is increased.

Table 2. *Standard thermodynamic functions of alcohols in water at 25 °C*

Property (units)	MeOH	EtOH	1-PrOH	2-PrOH	1-BuOH	2-BuOH	*i*-BuOH	*t*-BuOH	–CH_2–
ΔG^0_h (kJ mol^{-1})	−21.40	−20.98	−20.19	−19.90	−19.73	−19.15	−18.93	−18.89	0.74
ΔH^0_h (kJ mol^{-1})	−44.52	−52.40	−57.45	−58.21	−61.58	−62.72	−60.15	−63.92	−3.24
ΔS^0_h (JK^{-1} mol^{-1})	−77.5	−105.4	−125.0	−128.5	−140.4	−146.1	−138.3	−151	−13.3
ΔG^0_V (kJ mol^{-1})	4.48	6.3	8.8	6.9	11.7	9.4		7.0	
ΔH^0_V (kJ mol^{-1})	37.9	42.6	47.6	45.4	52.4	50.0	50.9	46.8	
ΔS^0_V (JK^{-1} mol^{-1})	112.5	121.5	130.2	129.3	136.4	133.9		133.4	
$C^0_{P,2}$ (JK^{-1} mol^{-1})	158	261	350	363	437	449	433	464	88
$C^*_{P,2}$ (JK^{-1} mol^{-1})	81.2	112.0	141.0	150.9	177.0	199	180	220	33
$C^0_{V,2}$ (JK^{-1} mol^{-1})	154	257	340	350	427	436	412	457	
$C^0_{P,g}$ (JK^{-1} mol^{-1})	43.9	65.4	87.1	88.7	110.0	113.3	111.3	113.4	
V^0_2 (cm^3 mol^{-1})	38.2	55.1	70.7	71.8	86.5	86.6	86.8	87.8	15.9
V^*_2 (cm^3 mol^{-1})	40.74	58.69	75.16	76.90	92.0	92.4	92.9	94.9	16.7
$10^2\ E^0_2$ (cm^3 K^{-1} mol^{-1})	1.3	1.6	3.2	4.1	(3.1)	(4.0)	(6.1)	(2.8)	(2.3)
$10^2\ E^*_2$ (cm^3 K^{-1} mol^{-1})	4.9	6.4	7.5	8.0	8.5	9.4	9.4	11.0	1.0
$10^4\ K^0_{S,2}$ (cm^3 bar^{-1} mol^{-1})	12.5	9.9	6.1	6.1	4.5	(3.5)	2.6	3.8	(−1.8)
$10^4\ K^*_{S,2}$ (cm^3 bar^{-1} mol^{-1})	43.0	57.7	64.7	77				99	20
$10^4\ K^0_2$ (cm^3 bar^{-1} mol^{-1})	12.8	10.2	6.9	7.2	5.2	(4.3)	4.3	4.3	(−2.4)
$10^4\ K^*_2$ (cm^3 bar^{-1} mol^{-1})	51.8	68.3	76.5						

Values in parenthesis have a large uncertainty.

There is a contraction in volume accompanying the transfer of the alcohol molecule from the pure liquid state to infinite dilution in water. This is again consistent with the so-called hydrophobic hydration which occurs with an economy of space. Since such transfer leads to a loss of free volume, the limiting partial expansibilities and compressibilities of the alcohols are anomalously low.

The variation of some of these quantities with temperature is shown in figure 7 for tert.-BuOH. As the temperature increases, the partial molar quantities tend to the corresponding molar quantities of the pure alcohols, indicating that the anomalous solute–solvent interactions are maximized at low temperatures.

The various models and theories that have been advanced to interpret the hydration thermodynamics of alcohols will be discussed in section 6.

The second virial coefficients y_{22} in eqn (4.10) which measure the solute pair interactions can be readily derived from the concentration dependence of the apparent molar quantities. Pair interaction parameters for two different solutes, 2 and 3, can also be derived through transfer functions. These interaction parameters for alcohols in water have been studied systematically by many authors. [**24, 29, 45, 47–49, 71, 73, 78, 109–111**]. The pair interaction parameters are summarized in table 3 for some of the alcohols in water. Where cross-parameters y_{23} have been measured in ternary aqueous solutions [**47**], they were found to lie close to the mean of the corresponding

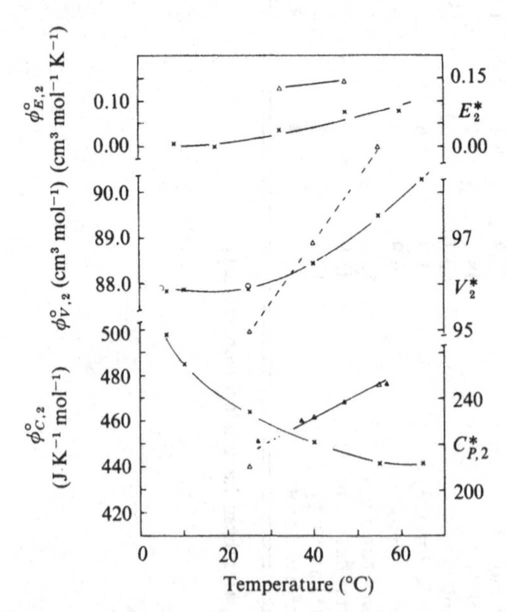

Figure 7. Infinite dilution apparent molar volumes, expansibilities and heat capacities (○, ×), and the corresponding molar quantities (△, ▲) as a function of temperature. Reproduced, with permission, from reference 55.

Table 3. *Pair interaction parameters for alcohols in water at 25 °C*

Parameter	MeOH	EtOH	1-PrOH	2-PrOH	1-BuOH	*t*-BuOH
g_{22}(J Kg mol^{-2})	−238	−230	—	−326	−1170	−607
h_{22}(J Kg mol^{-2})	218	248	540	—	1220	680
$c_{P,22}$(JK^{-1} Kg mol^{-2})	−3.2	−0.7	6.0	2.4	13.7	7.9
v_{22}(cm^3 Kg mol^{-2})	−0.13	−0.51	−1.06	−1.07	−1.35	−1.89
10^4 $k_{S,22}$ (cm^3 bar^{-1} Kg mol^{-2})	−0.31	−2.2	—	−3.1	—	−14.4

y_{22} and y_{33}, and they are therefore not included in table 3. At this point, it should be emphasized that for interactions between alkanols and their corresponding polyols, e.g. EtOH+ethane diol, PrOH+glycerol, BuOH+erythritol, the y_{23} do not fall within y_{22} and y_{33}.[**111**]

As expected, all the y_{22} parameters increase in magnitude with the size of the alcohol. The signs of g_{22} and h_{22} are opposite to those of the corresponding hydration functions which has led many authors[**112**] to interpret these parameters in terms of the formation of a quasi-dimer which is less solvated than two monomers. However v_{22}, $k_{S,22}$ and $c_{P,22}$ all have the *same* signs as the corresponding hydration functions, a result which is not consistent with the model of a contact pair. A solvent shared aggregation, would account better for the observed trends.[**20, 47, 110**] A question arises about the importance of the OH contribution to the total solute–solute interactions. Wood *et al.*[**109, 113**] have attempted to estimate the functional group contributions to the second virial coefficients, and, although interactions such as $CH_2 \cdots OH$ turn out to be large, they do not cause a change of sign of the $CH_2 \cdots CH_2$ contributions to g_{22} and h_{22}.

An example of the temperature dependence of the pair interaction parameters is shown in figure 8. The trends well illustrate the enthalpy–entropy compensation for these interactions, with the sign of g_{22} determined by Ts_{22} rather than h_{22}. The main contribution to the negative g_{22} may possibly not be due to the structural hydration interaction; alternative explanations have been suggested.[**114–116**] However, it will be recalled that the scattering data are quite compatible with the thermodynamic results, showing that at least part of the resultant free energy is of a structural origin. Furthermore, the hydration functions in table 2, as well as the signs and magnitudes of g_{22} and h_{22} resemble those of hydrocarbons in aqueous solution.[**64, 117**] In contrast, g_{22} values for polyhydric alcohols, e.g. glycerol, mannitol, are found to be small but positive[**118**], with $|h_{22}| > T|s_{22}|$. In any discussion of hydration functions and pair interaction parameters it must be stressed that the large specific heat effects are indicative of a high degree of temperature sensitivity and that reversals in the signs of some of the thermodynamic quantities are to be expected at low temperatures. For example, g_{22} of

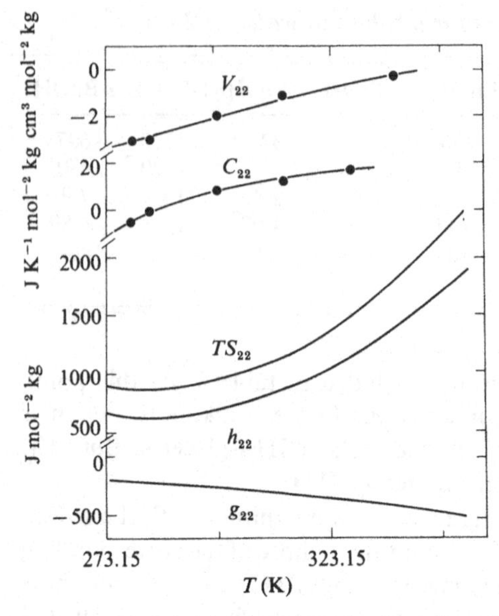

Figure 8. Second virial coefficients for the various thermodynamic functions of tert.-BuOH in water as a function of temperature. Reproduced, with permission, from reference 49.

alkyldimethylamine oxides are positive at 0 °C[**119**], and g_{22} of alcohols might also be expected to become positive below that temperature[**49**]. The interpretation of the interaction parameters and especially their higher temperature and pressure derivatives remains complex and provides a severe test for any quantitative theory.

The complexities of solute–solute interactions at high concentrations are best seen from plots of the partial molar quantities. The partial molar volumes of the three first normal alcohols normalized to V_2^0 are shown in figure 9. Minima are observed in the water-rich region which occur at lower mol fractions and are sharper the longer the alkyl chain length. It is interesting to note that *n*-BuOH becomes completely miscible with water in the presence of a small quantity of surfactant such as sodium dodecylsulphate and its partial molar quantities then fall quite in line with the other alcohols.[**120**] Molar mixing functions, based as they are on ideality at both ends of the composition range, tend to hide these changes in the water-rich region. The sharpness of the transitions in the water-rich region are even more obvious when the various partial molar quantities of tert.-BuOH at 298 K are compared, as in figure 10. All the abrupt changes occur near $x_2 \simeq 0.05$ and normal behaviour is established $x_2 > 0.12$, where the partial molar quantities tend to constant values. These trends are amazingly similar to those of surfactants in aqueous solution, except that the changes are then at much lower concentrations and even more abrupt. For $x_2 < 0.05$ tert.-BuOH

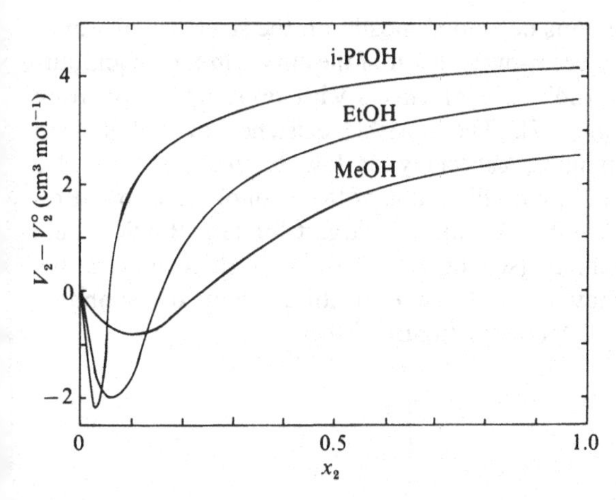

Figure 9. Partial molar volumes minus the standard infinite dilution value of alcohols in water at 25 °C as a function of the mol fraction of the alcohols.

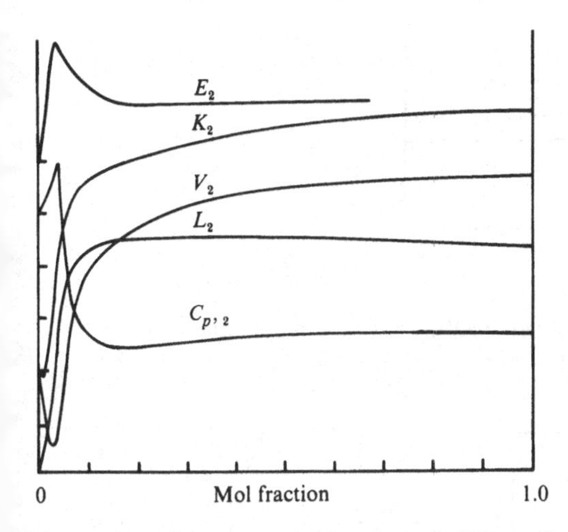

Figure 10. Partial molar quantities of tert.-BuOH as a function of its mol fraction at 25 °C.

molecules are in an aqueous environment and appear to interact strongly with each other. Beyond $x_2 \simeq 0.13$ the added alcohols experience essentially other alcohol molecules, so that $Y_2 \simeq Y_2^*$. The alcohol–water mixtures therefore undergo some higher order microphase transition at $x_2 \approx 0.05$. Similar changes are also apparent in the energy–volume coefficients defined by eqn (4.14).

The temperature dependence of $\phi_{C_p,2}(x_2)$ is shown for tert.-BuOH in figure 11. As with all properties except free energies, the changes in the thermodynamic quantities are larger and sharper at low temperatures, as we might expect if the interactions depended mostly on the structure of water. A remarkable feature is the cross-over region of the curves in the neighbourhood of $x_2 = 0.1$. This is indicative of effects with opposite temperature dependences (see also figure 7). The possible existence of well-defined liquid-state molecular domains, especially at low temperatures, is also suggested by the solid–liquid phase diagrams of the alcohol–water mixtures (see for example reference 100). The effect is largest for tert.-BuOH where a hydrate actually crystallizes (see figure 12(a)), and there is also the indication of peritectic behaviour in the concentration region corresponding to the clathrate hydrate stoichiometry (figure 12(b)).

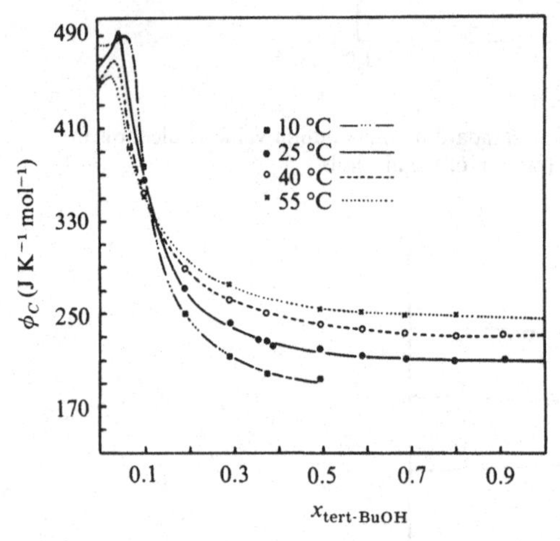

Figure 11. Apparent molar volume of tert.-BuOH in water at different temperatures. Reproduced, with permission, from reference 55.

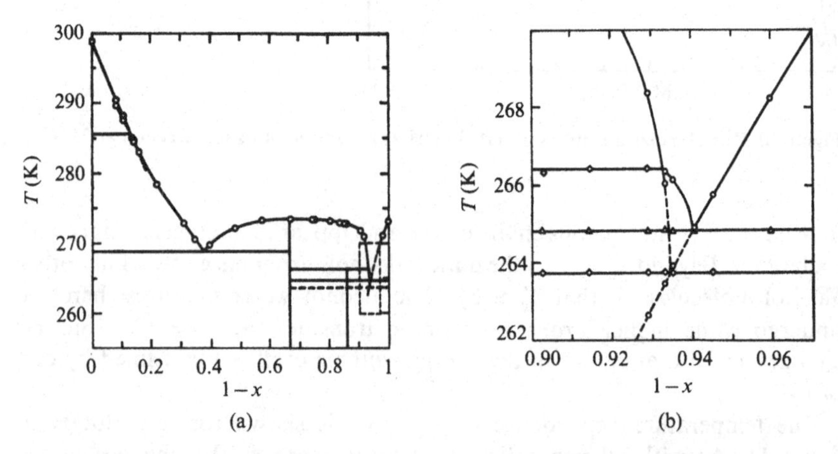

Figure 12. Solid–liquid diagram for $(1-x)H_2O + x$ tert.-BuOH (a) and expanded view of the region of incongruent-compound formation (b). Reproduced, with permission, from reference 100.

The high-temperature behaviour of alcohol–water mixtures is more normal. All butanols that are not completely miscible at room temperature become homogeneous at high temperatures at an upper consolute temperature. The liquid–liquid phase equilibria at high pressure are more interesting, as seen in figure 13 for 1-BuOH. Pressure increases the miscibility, in agreement with the negative volume of mixing. Therefore the nonideality of the system decreases with both temperature and pressure.

Most of the available experimental results relate to ambient and higher temperatures. The volume measurements on undercooled solutions reported recently by Sorensen are therefore of particular interest.[95] The V–T–x data for tert.-BuOH–water mixtures are shown in figure 14. The temperature of maximum density is seen to shift to lower values as x_2 increases and the

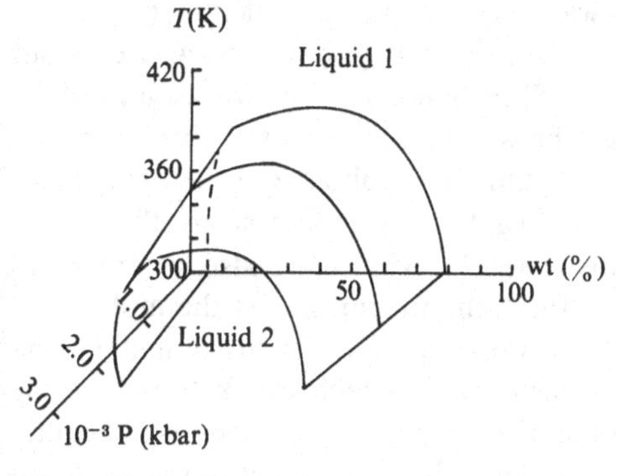

Figure 13. Liquid-vapour diagram of n-BuOH in water.

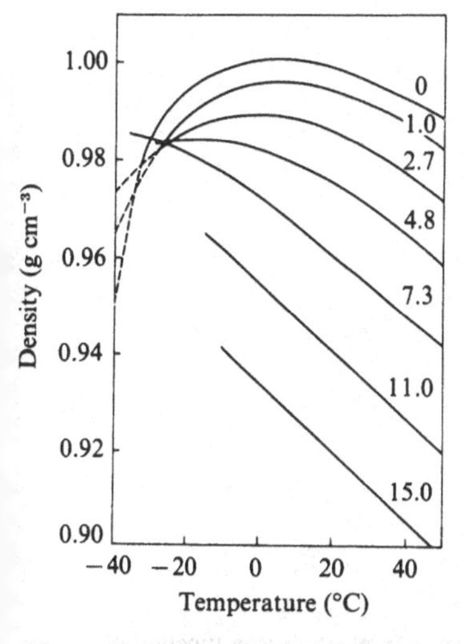

Figure 14. Densities of tert.-BuOH–water mixtures as a function of temperature (numbers refer to mol per cent tert.-BuOH). Broken lines indicate extrapolated densities to −40 °C. Reproduced, with permission, from reference 95.

density curves exhibit a complex cross-over region, i.e. as the alcohol which has a lower density than water is added to water, so the mixture becomes denser. At $x_2 \approx 0.05$, θ seems to disappear altogether. Although the data are hardly precise enough for the calculation of $V_2(x_2)$, and some values are even based on extrapolated densities, the author states that the characteristic V_2 minimum (see figure 9) disappears below -30 °C. By comparison, the V–T–x_2 behaviour of N_2H_4–water mixtures shows none of the above anomalies at subzero temperatures.

The partial molar quantities of the alcohols at high x_2 are not very informative, since the interactions are distributed over a large number of molecules. It is therefore preferable to examine the properties of water in such mixtures. Little is known about such properties. The values of V_1^0 are lower than the molar volume of water, as expected, since the characteristic structure of liquid water does not exist when water is surrounded by alcohol molecules. Some of the trends with temperature are surprising, however, as seen from the data of Sakurai and Nakagawa [**75**] in figure 15. The low values of V_1^0 and E_1^0 for the lower alcohols may be attributable to extensive hydrogen bonding between the water molecules and the alcohols. On the other hand E_1^0 in MeOH and tert.-BuOH are negative, in contrast to the values corresponding to the other alcohols. The $C_{P,1}^0$ values are also surprising in that they remain larger than, or of the same magnitude as the molar heat capacity of pure water. If the high value of $C_{P,1}^*$ is attributed to the three-dimensional network of hydrogen bonds which breaks down as the temperature is increased, one would then expect $C_{P,1}^0$ to be significantly smaller, as in the case of the volumes. This phenomenon is general and has been observed in many polar organic solvents. For example, such behaviour

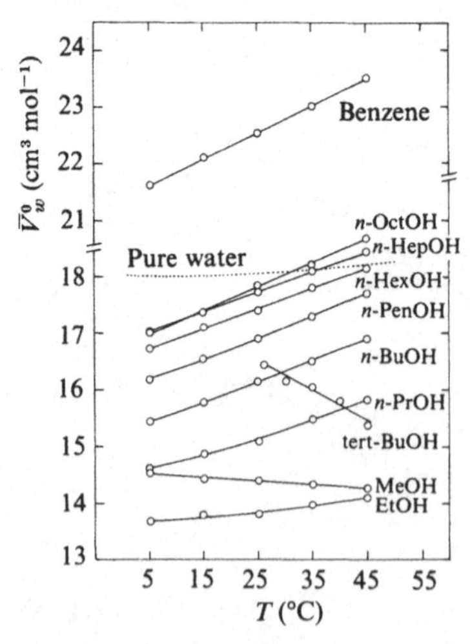

Figure 15. Infinite dilution partial molar volumes of water in alcohols as a function of temperature. Reproduced, with permission, from reference 75.

also occurs in alkoxyethanols, shown in figure 16. A possible explanation for these high heat capacities has been offered by DeGrandpré *et al.*[**22**], based on their investigation of the near-infrared spectra of water in alcohol–water mixtures, where they showed that the water molecules in an excess of alcohol are extensively hydrogen-bonded to the alcohols. More work on the properties of water in alcohols and other non-aqueous solvents is certainly warranted.

4.3. *Dynamic Properties*

'The structural properties of a liquid are reflected by the nature of the molecular motions in the liquid'. This statement by Goldammer and Hertz[**123**] will serve as a guideline for the following discussion of the microscopic and macroscopic transport properties of alcohol–water mixtures. Much of our available information derives from the extensive nuclear magnetic relaxation measurements by Hertz and his colleagues[**123–126**], designed to investigate the dynamic aspects of the dilute solution anomalies which are so evident in the scattering and thermodynamic properties. By a combination of ^{1}H and ^{2}H (D) relaxation measurements, it is possible to determine the rotational and translational diffusion rates of both molecular species in a binary mixture. More detailed information becomes available from supplementary measurements with other nuclei, e.g. ^{13}C, ^{17}O, or ^{15}N. Where the organic solute does not possess exchangeable protons, the detailed dynamics of both solvent and solute can be completely resolved. In the case of alcohols this is unfortunately impossible because the –OH protons are subject to fast exchange with water protons. ^{1}H and ^{2}H relaxation measurements can then yield only averaged rates of motion. ^{17}O measurements on water, although more difficult to perform, provide the additional information. The measured proton spin-lattice relaxation rate T_1^{-1} is composed of an intramolecular and an intermolecular part:

$$T_1^{-1} = T_1^{-1}{}_{\text{intra}} + T_1^{-1}{}_{\text{inter}}. \qquad (4.15)$$

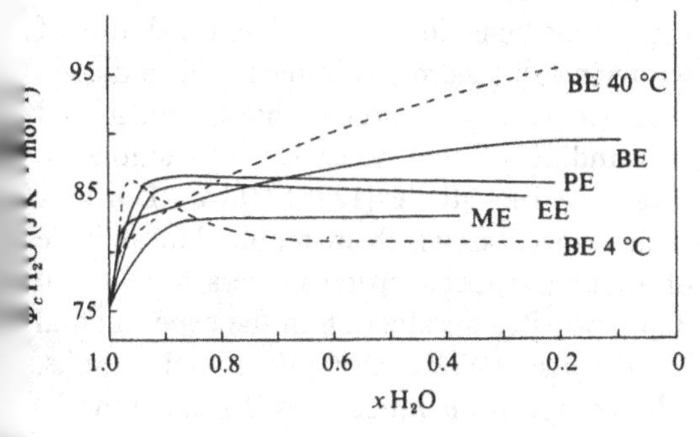

Figure 16. Apparent molar heat capacities of water in 2-alkoxybutanols–water mixtures at 5 °C, and 4° and 40 °C in the case of 2-butoxyethanol. Reproduced, with permission, from reference 121.

For deuterons, the latter contribution is zero, because the relaxation arises almost completely from electric quadrupole field gradient effects, whilst proton relaxation is due to magnetic dipole–dipole interactions. The first term on the right hand side of eqn (4.15) can thus be obtained from the deuteron T_1^{-1}. The rotational tumbling time τ_r of a molecule is related to $T_1^{-1}{}_{intra}$ by

$$T_1^{-1}{}_{intra} = F\tau_r \tag{4.16}$$

where the factor F can be calculated from the geometry of the molecule. It should also be noted that the interaction between protons and deuterons is very much weaker than that between protons. The experimental evaluation of τ_r of both components in a binary mixture in principle involves measurement of ^1H and ^2H relaxation rates in solutions of various fixed compositions where the $^1H/^2H$ ratio of one component at a time is altered. The measurements are then extrapolated to obtain infinite dilution values. Such experiments are extremely laborious and the extrapolations somewhat uncertain. Goldammer and Zeidler introduced certain simplifications based on the assumption that

$$\tau_r^D = k\tau_r^H$$

where k is a constant. The validity of this relationship has been tested. $T_1^{-1}{}_{intra}$ for the solute can then readily be determined from

$$T_1^{-1}{}_{intra}(^1H) = LT_1^{-1}(^2H)$$

where $L = T_1^{-1}(^1H)/T_1^{-1}(^2H)$ as $x_2 \rightarrow 0$. The corresponding information for water can only be obtained (at $x_2 \rightarrow 1$) where no solute–water proton exchange is possible, i.e. not for alcohol–water mixtures. However, the physical properties of aqueous solutions of aprotic molecules, such as tetrahydrofuran (THF) and acetone, bear a very close resemblance to those of alcohols [**64**] so that we consider it likely that the diffusional properties of the water molecules in alcohol–water mixtures can be inferred from those observed for aqueous THF and acetone solutions.

An analysis of the available experimental data leads to the important conclusion, that where some anomalous behaviour can be identified, it is of a much lesser degree than the scattering and thermodynamic results indicate. Nevertheless, the nuclear spin-lattice relaxation studies have contributed markedly to our general understanding of the molecular interactions in aqueous solutions. A disadvantage of earlier studies [**123, 124**] was their lack of precision, especially with the dilute solutions which are required for reliable extrapolations. The advent of more advanced spectrometers and signal accumulation techniques has made possible a reduction in the experimental uncertainty of T_1^{-1} to $\pm 3\%$ [**120**]; also (Wells, Derbyshire and Franks, unpublished results), so that the composition range $x_2 \leqslant 0.1$ can now be investigated with some confidence.

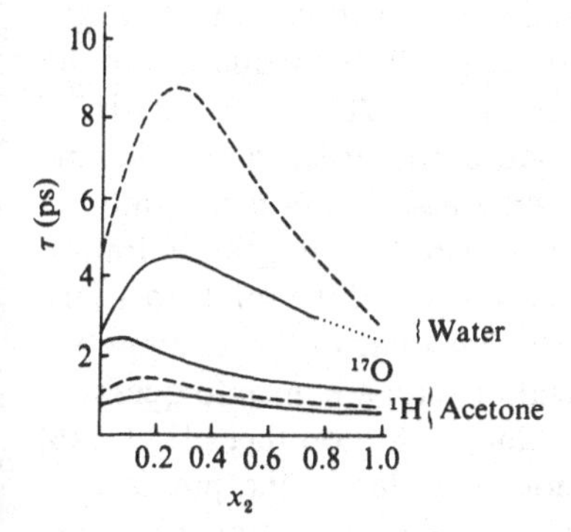

Figure 17. Proton and ^{17}O n.m.r. correlation times τ in acetone–water mixtures at 5 °C (broken line) and 25 °C (solid line); extrapolated values are shown by dotted lines. After Goldammer & Hertz.[**123**]

The most striking results of the relaxation studies are summarized in figure 17. The system acetone–water has been chosen because the information is more complete than that which can be extracted from the experiments on alcohol–water mixtures, for reasons mentioned above. The three lower curves describe the rotational diffusion of acetone and the two upper ones the corresponding motions of water. In the limit of $x_2 \rightarrow 0$, the acetone molecules are able to tumble more rapidly than the much smaller water molecules. With increasing acetone concentration, the motions of water molecules are seen to suffer a severe inhibition, especially at 5 °C, whereas the motion of the solute molecules is hardly affected. The small discrepancy between the ^{1}H and ^{17}O correlation times suggests that the tumbling of acetone is subject to some degree of anisotropy.

The temperature dependence of τ_r, as measured for the two molecular species, also emphasizes the complex solution behaviour. Thus, τ_r for H_2O shows a marked temperature dependence which is almost completely lacking in τ_r of acetone. Furthermore, the temperature coefficient of τ_r (H_2O) is very sensitive to the solute concentration, with a maximum near $x_2 = 0.25$. A qualitatively similar behaviour has been established for the concentration and temperature dependences of the (translational) self-diffusion coefficients of the two molecular species in the mixtures.

In the case of alcohol–water mixtures only averaged τ_r values can be obtained for the alcohol hydroxyl and water protons. The indications are that here too the diffusional motions of water are considerably slower than those of the larger alcohol molecules (e.g. tert.-BuOH) and that the rotational tumbling times, as obtained from T_1 measurements on several different nuclei, show the same dependences on concentration and temperature as those depicted in figure 17. [**123, 124**]

The $\tau_r(x_2, T)$ behaviour so far described reflects intramolecular motions and solute–water interactions. Solute aggregation, such as suggested by the light scattering and SAXRS data, should become evident from an analysis of the intermolecular contribution to T_1^{-1}. Leiter *et al.*,[**126**] on the basis of measurements on aqueous solutions of acetone, acetonitrile, trimethylamine and tetramethylurea, have concluded that such association could only be detected for the last mentioned solute. It seems, therefore, that concentration fluctuations do not affect the microdynamic behaviour of the solute to a very marked extent, although the possibility exists that some of the assumptions made in the interpretation of the n.m.r relaxation data are open to question, e.g. the assumed shapes and dimensions of solute hydration shells.[**123, 126**]

The molecular details of the self-diffusion of water in the presence of tert.-BuOH have also been investigated by quasi-elastic neutron scattering [**127**]. The analysis of the data is indicative of a diffusion mechanism, described by the authors as translational 'shuffling', rather than proton jump diffusion, with an occasional rotational diffusive step. In harmony with the n.m.r. results, neutron scattering data are not consistent with a picture of long-lived hydration shells. The inelastic portions of the spectra indicate that tert.-BuOH does indeed enhance the spatial organization of the water molecules, but such 'structures' bear no relation to ice.

The macroscopic transport properties of alcohol–water mixtures, in particular the viscosity, have been extensively studied, but few precision measurements on dilute solutions are on record. Herskovits and Kelly[**128**] investigated the role of alcohols as protein denaturants with particular reference to the hydrodynamic aspects. A popular way of interpreting viscosity/concentration relationships is in terms of an extended Einstein equation, e.g. of the type

$$\eta/\eta_0 = 1+2.5\Phi+C'\Phi^2 \tag{4.17}$$

where Φ is the volume fraction and the coefficient C' represents the contribution of solute–solute interactions (between spherical particles) to the relative viscosity. A modified form of eqn (4.17), the so-called Jones–Dole equation,[**129**] expresses η/η_0 in terms of the molal solute concentration:

$$\eta/\eta_0 = 1+Bm+Cm^2. \tag{4.18}$$

This equation has been used extensively to probe hydration effects in aqueous electrolyte solutions.[**130, 131**]. The coefficient B is clearly related to the factor 2.5 in eqn (4.17), but once hydration is invoked, it then requires a number of assumptions and empirical 'corrections', because the hydrodynamic flow unit may not correspond to the bare particle (solute molecule). Holtzer and Emerson[**131**] correctly stated that the intrinsic viscosity

$$[\eta] = \left(\frac{\eta-\eta_0}{\eta}\right)_{m\to 0}$$

is the correct quantity which should be used to calculate B. Despite the many reservations involved in the use of eqns (4.17) and (4.18) to model the flow behaviour of solutions of molecules (rather than large non-interacting spherical particles), correlations have been established between the viscosity increment and various molecular properties of members of homologous series. For instance, the observed increase in B for each additional CH_2 group is quite constant, at 16–18 $cm^3\ mol^{-1}$, for the alcohol and amide series. This is 5–6 cm^3 in excess of the predicted Einstein value of 2.5 V_h, where V_h is the effective molar volume of the hydrodynamic flow unit. It seems impossible to rationalize the B-coefficient data for alcohols, amides, urea derivatives and polyhydroxy compounds without recourse to assumptions about water structure promotion/breaking, but this is exactly the information which the viscosity measurements are meant to provide.

On a qualitative basis, viscosities provide some interesting indications about solution behaviour. Alcohol–water mixtures exhibit extrema with positions which depend on the molecular weight and the temperature and occur at the same concentrations as the scattering maxima (figure 6). Even more informative is the shape of $d\eta/dx_2$, shown for the systems EtOH–water and dioxan–water in figure 18[**132**]. Both systems exhibit viscosity maxima of similar magnitudes, but only the EtOH–water mixtures provide evidence of two (or more?) competing effects which balance at $x_2 = 0.08$ which also happens to be the concentration corresponding to the extrema in $V_2(x_2)$, $C_{P,2}(x_2)$ and $E_2(x_2)$.

4.4. *Some Other Physical Properties*

There is little to be gained from an exhaustive survey of the physical properties of alcohol–water mixtures. Suffice it to note that most of the properties exhibit extrema or inflexions at the characteristic concentrations which have been

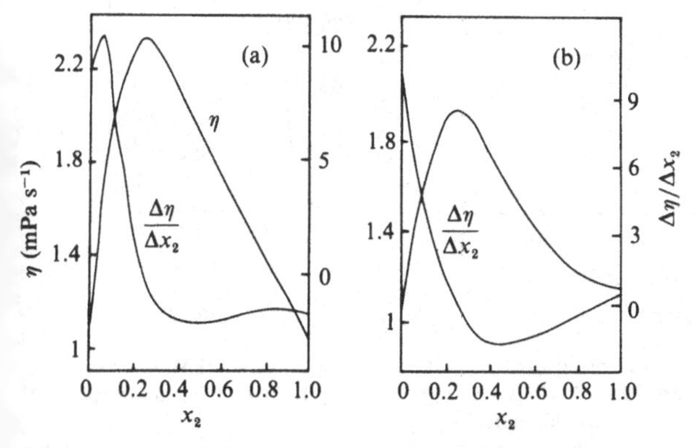

Figure 18. Viscosity η and incremental viscosity, $\Delta\eta/\Delta x_2$, as functions of solute concentration at 25 °C for (a) EtOH–water and (b) dioxan–water mixtures. After Kay.[**132**]

identified in sections 4.1–4.3. We shall therefore confine ourselves to a few pertinent examples taken from a range of physical techniques.

The far-ultraviolet absorption spectra of alcohols are more intense than that of water, so that their absorbances in aqueous solution can readily be measured in the 54 000 cm^{-1} region. The molar absorbances are well established and it has been shown that in the dilute solution range they do not obey Beer's law[**133**] but exhibit extrema.

The acoustic properties (sound propagation, absorption and dispersion) of alcohol–water mixtures have been subjected to detailed investigation over a wide frequency range, mainly by Blandamer and his colleagues.[**134**] Most measurements have been performed in the range 1–250 MHz because this allows the study of molecular relaxation processes with relaxation times in the range $10^{-6} < \tau < 10^{-9}$ s. Ultrasound absorption is usually expressed as (α/ν^2), where α is the attenuation coefficient. The frequency-dependent contribution to (α/ν^2) is said to derive from the perturbation of some simple chemical equilibrium, e.g. between two conformational states of a molecule. Alcohol–water mixtures give rise to complex absorption spectra; some of the salient features are shown in figures 19 and 20. At low x_2 values α/ν^2 appears to be independent of x_2, reminiscent of the light scattering curves in figure 5. According to figure 19, for a given alcohol concentration α/ν^2

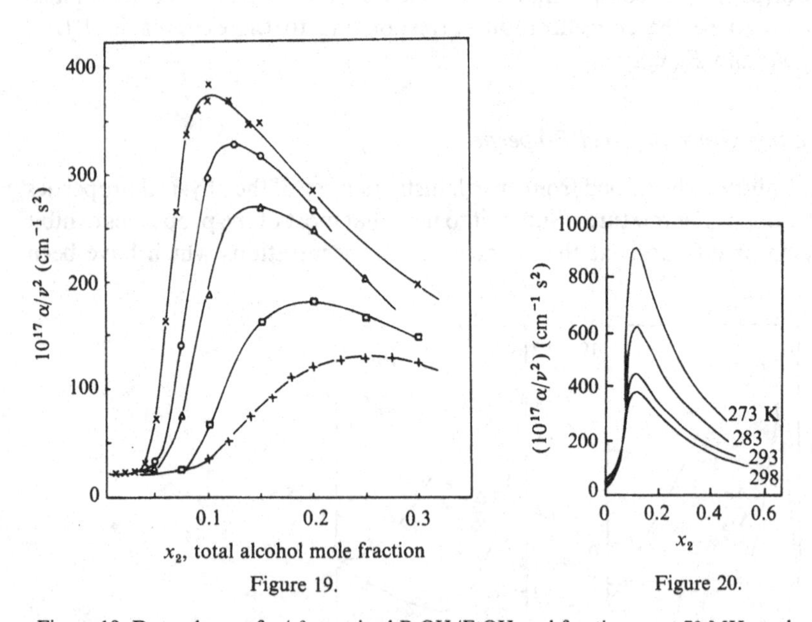

Figure 19. Dependence of α/ν^2 on mixed PrOH/EtOH mol fraction x_2 at 70 MHz and 298 K. PrOH: EtOH ratios are 100:0 (×); 70:30 (○); 50:50 (△); 20:80 (□) and 0:100 (+). Reproduced, with permission, from Blandamer.[**134**]

Figure 20. Dependence of α/ν^2 (at 70 MHz) of *n*-PrOH–water mixtures on temperature. After Blandamer.[**134**]

increases with the number of methylene groups, as does also the sharpness of the curve; the position of the maximum shifts to lower x_2 values as the ratio n-PrOH/EtOH increases at a constant total solute concentration.

The temperature dependence of α/ν^2 is of particular interest: it is opposite in sign to that of the light scattering and SAXRS intensities, i.e. whatever the process that is responsible for the absorption maximum, its intensity decreases with rising temperature, except at low concentrations, where the temperature dependence changes sign. The sharp increase in α/ν^2 over a narrow concentration range can therefore not be the result of the concentration fluctuations which give rise to the observed scattering maxima. The exact origin of the α/ν^2 behaviour is still uncertain.

As final example of an eccentric concentration dependence of a physical property of alcohol–water mixtures, we cite the n.m.r. chemical shift of ^{129}Xe. The shift is large and especially sensitive to the physical environment of the nucleus. It has its origin mainly in van der Waals interactions between Xe and its surroundings[**135**], although the shift in water is larger by 45 p.p.m. than would be predicted on the basis of van der Waals interactions! The excess shift in water is well accounted for in terms of the clathrate-like nature of the Xe hydration shell and is fully compatible with the Xe shift in the crystalline clathrate hydrate.[**136**] Presumably the effect of a solute on the Xe shift must reflect the influence of that solute on the integrity of the solvation shell. Figure 21 compares the effects of a number of solutes on the Xe shifts.[**137**] The now familiar pattern emerges: alcohols give rise to extrema the sharpness of which increases with increasing alkyl chain length. By comparison, the excess shifts in solutions of dioxan and acetonitrile are much smaller and do not show such a pronounced concentration dependence over a limited concentration range.

Curiously, some properties of water which might be expected to be highly

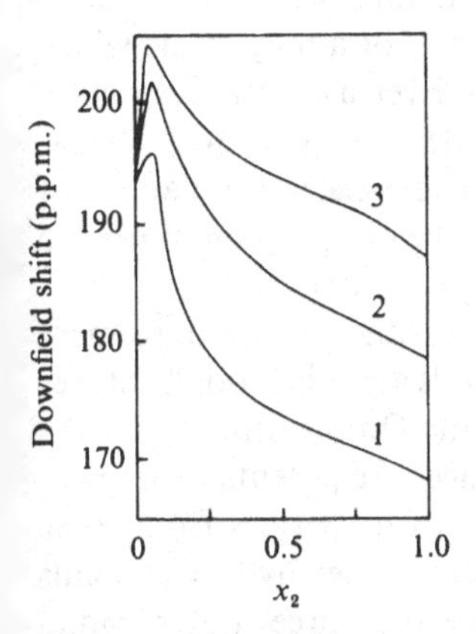

Figure 21. Chemical shifts of Xe in aqueous mixtures of n-PrOH (1), 2-PrOH (2) and tert.-BuOH (3), referred to the chemical shift of Xe gas.[**137**]

sensitive to the presence of alcohols do not exhibit such sensitivity. An example is provided by the O–D stretching vibration, as measured in dilute solutions of D_2O in H_2O. The addition of alcohols produces only very minor shifts in the frequency [**133**], in the same direction as would be produced by moderate elevation in the temperature.

5. Computer Modelling

Reference has already been made to recent *ab initio* calculations of alcohol–water pair potential functions, [**21, 22**] The geometries of the molecules were derived from microwave measurements and the molecular orbital (MO) calculations were performed on 500 dimer configurations.

The resulting dimer potential energy curves, shown in figures 22(a) and (b), for MeOH–H_2O and tert.-BuOH–H_2O, exhibit marked anisotropy. The two deep wells correspond to the hydrogen bond positions in which the alcohol –OH serves as proton donor and acceptor respectively. The corresponding hydrogen bond energies are summarized and compared to earlier data in table 4. In contrast to a narrow, deep well in the potential surface corresponding to the hydrogen bond configuration, the minimum corresponding to the HO–$CH_3 \cdots$ water configuration is shallow. Figures 23(a) and (b) show the potential energy curves for the MeOH–water and tert.-BuOH–water dimers with water placed in the direction O–alkyl. The shapes of the latter suggest that water molecules in the hydrophobic region are influenced primarily by other water molecules and that water–alcohol interactions therefore do not contribute significantly to the macroscopic solution properties except in the narrow regions corresponding to the hydrogen bond directions (figure 22). Thus, models earlier proposed [**20, 64**] are consistent with the pair potentials and the observed behaviour of dilute solutions can be accounted for largely by hydrophobic hydration. The consideration of contributions from specific solute–water hydrogen bonding in thermodynamic interpretations, though correct in principle, may well introduce uncertainties of a magnitude which cancel any potential benefit from such a refinement. Even a cursory inspection of the calculated hydrogen bond energies in table 4 lends support to this view.

The pair potentials have formed the basis of Monte-Carlo simulations of infinitely dilute alcohol solutions. [**21, 22**] Actually the samples contained 215 water molecules and one alcohol molecule (i.e. $x_2 = 0.005$).

Although the mixing process was found to be exothermic, in agreement with experiment, the potential energies of the alcohols are substantially higher in the aqueous solution than in the pure liquid state. On the other hand, the introduction of the alcohol molecules slightly reduces the potential energy of the aqueous component. Hence, the exothermic mixing derives from some change in the water structure rather than from newly formed hydrogen bonds between alcohol and water. Small increases in the three water radial distribution functions $g_{O-O}(r)$, $g_{O-H}(r)$ and $g_{H-H}(r)$ accompany the introduc-

Table 4. *Calculated hydrogen bond energies of alcohol–water dimers (kJ mol*$^{-1}$)

Alcohol	Proton donor	Proton acceptor	Reference
Methanol	−24.6	−17.2	21
	−26.2	−21.8	138
Ethanol	−39.7	−32.2	139
tert.-Butanol	−25.6	−21.7	22

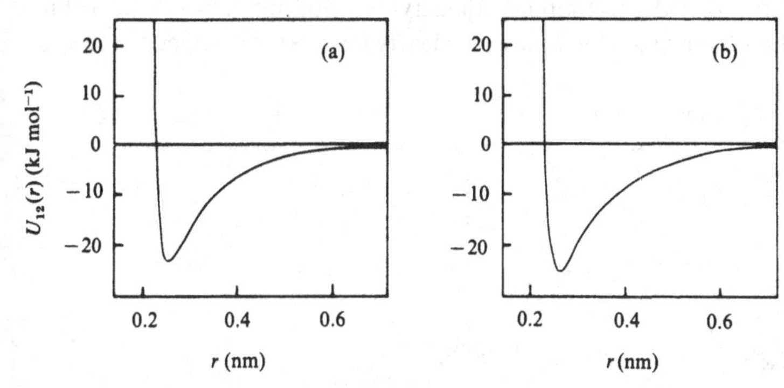

Figure 22. Potential energy curves for (a) MeOH–H_2O and (b) tert.-BuOH–H_2O dimers in the direction –OH–water; redrawn from references 21 and 22.

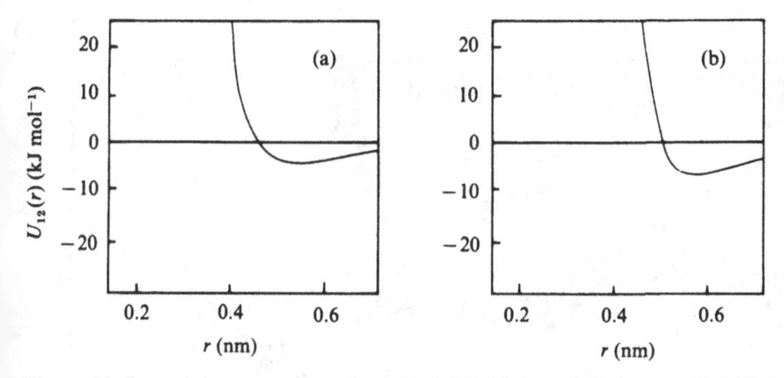

Figure 23. Potential energy curves for (a) MeOH–H_2O and (b) tert.–BuOH–H_2O dimers in the direction O–alkyl–water; redrawn from references 21 and 22.

tion of the alcohol molecule, reminiscent of the effects produced by the introduction of a hard sphere into water.[6]

As might be expected from the shapes of the water–alcohol potential functions, the solution structures exhibit marked anisotropy, depending on whether a water molecule is in the region where it can form a water–alcohol hydrogen bond or not. An analysis of the water–water interactions in the alcohol primary hydration shell (nearest neighbours) indicates that despite the repulsive alcohol–water interactions in the hydrophobic region, the potential energy is lowered compared to that in pure water, i.e. the phenomenon referred to as hydrophobic hydration. [**20**] An important conclusion from the computer model is therefore that hydrophobic hydration exists even when the solute molecule also interacts with water by hydrogen bonding, and that the general features of such hydration closely resemble those established for solutions of hard spheres [7] and hydrocarbons [**8, 9**] by various computer simulation methods; see also papers and discussion in reference 2.

The water molecule probability density distribution about the solute molecule, shown for MeOH in figure 24, clearly reveals the existence of several

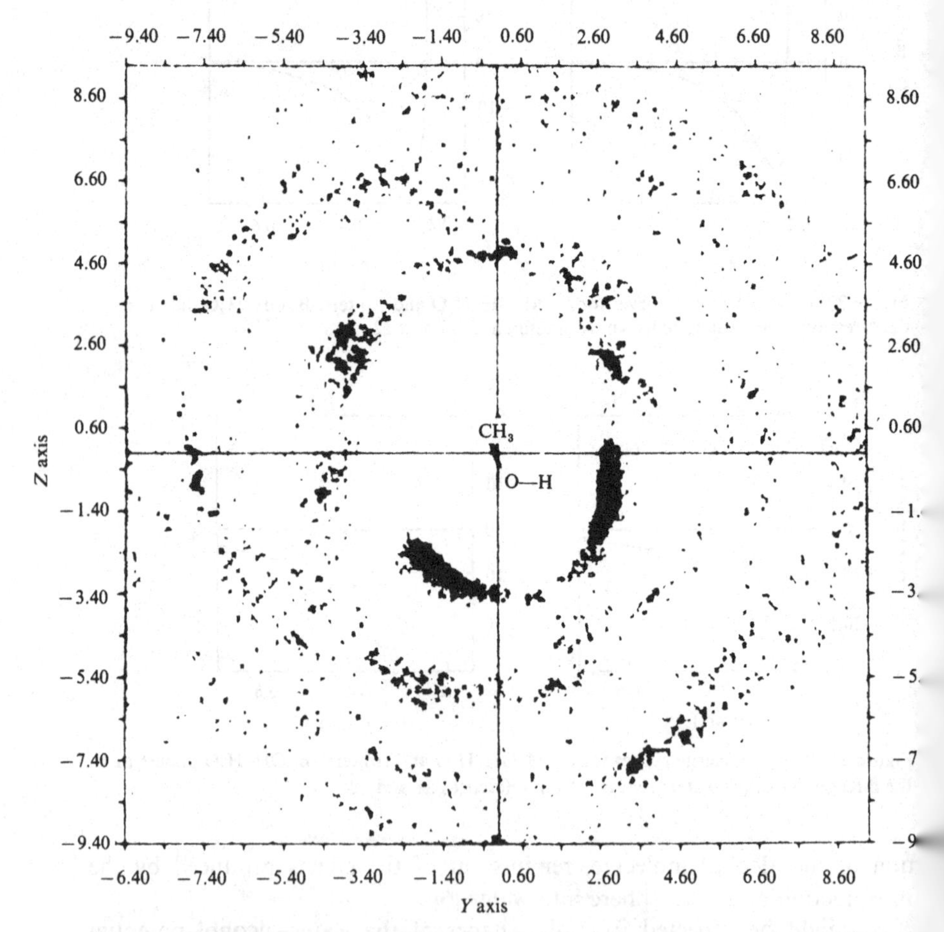

Figure 24. Water oxygen atom density distribution (diffusion averaged) around a MeOH molecule according to the Monte Carlo simulation of an infinitely dilute solution. Reproduced, with permission, from reference 21.

concentric regions of high density around the alkyl group. This pattern is even more pronounced for tert.-BuOH and is not found in the corresponding density distribution in pure water where the correlations with molecules beyond the nearest neighbours seem to be almost uniform. In the case of tert.-BuOH the solute influence on the surrounding aqueous component extends to a sphere of at least 20–25 water molecules which estimate agrees closely with that derived from the n.m.r. relaxation studies. [**123**]

While the above calculations are relevant to the infinitely dilute solutions, their extension to finite concentrations requires at least an alcohol dimer potential. Nakanishi and his colleagues have reported such potential functions for MeOH and tert.-BuOH dimers. [**140**, **141**] This has enabled them to extend their computer modelling studies to 'real' solutions of MeOH ($x_2 = 0.05$) and tert.-BuOH ($x_2 = 0.03$). While the former simulation was again of the Monte Carlo type, the latter study employed the molecular dynamics technique. It is now possible to compare the computed thermodynamic properties with experimental excess functions, e.g. ΔG_e, ΔH_e, discussed in section 4. The agreement is satisfactory. An analysis of the data indicates that the potential energy of the alcohol is not lowered (compared to the pure liquid alcohol) in the aqueous solution. On the other hand, the presence of the alcohol molecules lowers the potential energy of the solvent and also sharpens the three radial distribution functions pertaining to the solvent: $g_{\mathrm{O-O}}(r)$, $g_{\mathrm{O-H}}(r)$ and $g_{\mathrm{H-H}}(r)$; this effect is more pronounced in a 5 mol per cent solution than at infinite dilution.

Of greater interest are the alcohol–alcohol interactions. Here we distinguish between –OH correlations, indicative of solute–solute hydrogen bonding and Me–Me correlations resulting from hydrophobic interactions. Only the latter effect is apparent in the results: in MeOH $g_{\mathrm{Me-Me}}(r)$ exhibits a distinct peak at 0.38 nm, whereas alcohol–alcohol hydrogen bonding is negligible. At the concentration used in the simulation the average association number is 1.33, a result which is fully consistent with the partial molar properties. As might be expected, the effects are more pronounced for tert.-BuOH.

The $g_{\mathrm{Me-Me}}(r)$ function exhibits a second peak at 0.7 nm which can be identified with the solvent separated association of MeOH molecules. This type of interaction was first suggested ten years ago [**20**] and is at variance with the classical model of the hydrophobic interaction, based on a contact interaction between alkyl residues [**112**, **142–144**], which has become popular with biochemists. At first sight such a distinction may appear trivial, but it has important implications as regards the range of the hydrophobic interaction. This is illustrated in figure 25 which shows the potential of mean force $W_{\mathrm{MeOH-MeOH}}(r)$. The interaction is seen to be complex, long range and with the second (0.7 nm) minimum much deeper than the primary minimum at contact distance. This result is compatible with the calculations on pairs of methane molecules in aqueous solution [**8**, **9**] and with the well-established crystal structures of clathrate hydrates.

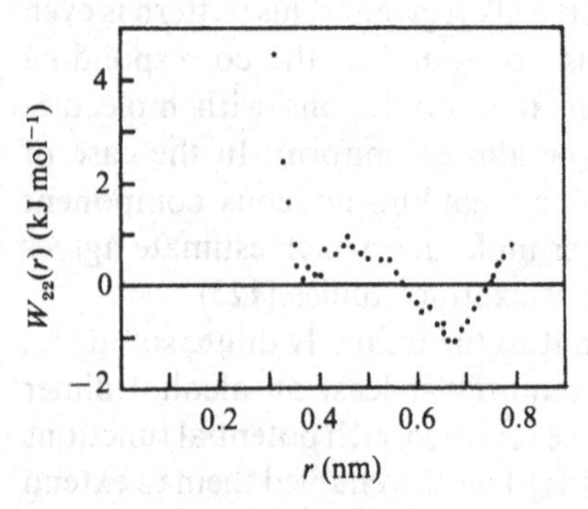

Figure 25. Potential of mean force for the MeOH dimer in aqueous solution ($x_2 = 0.05$) at 298.15 K, calculated from a Monte Carlo computer simulation.[140]

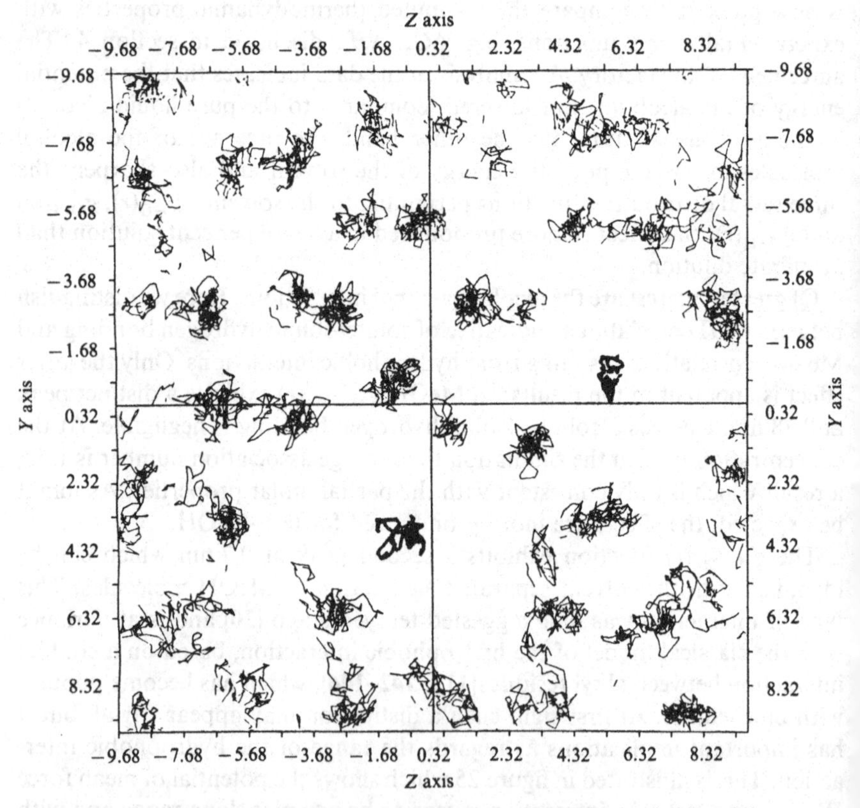

Figure 26. Two-dimensional trajectories of two tert.-BuOH molecules (bold) and the surrounding water molecules, taken over a period of 10 ps after the attainment of equilibrium, according to the Molecular Dynamics simulation of a 3 mol per cent tert.-BuOH solution. Reproduced, with permission, from reference 141.

The molecular dynamics simulation of a tert.-BuOH–water mixture also permits the estimation of the motional properties of the molecular species. Thus, figure 26 shows the (two-dimensional) 'trajectories' of a pair of tert.-BuOH molecules and the surrounding water molecules over a period of 10 ps[**141**]. The water-separated interaction is again clearly indicated as is also the localization of the molecular motions about mean positions corresponding to a quasi-clathrate geometry. The trajectories of the water molecules are quite tight in accord with the n.m.r. relaxation[**123, 124**] and the quasi-elastic neutron scattering results[**127**] already referred to, and as suggested earlier by Geiger *et al.*[**7**] on the basis of their own molecular dynamics simulation of a rather more artificial aqueous solution.

In summary, the results of the computer calculations are fully consistent with all the experimental results on dilute solutions described in section 4. They are particularly valuable because they provide an approximate shape of $W_{22}(r)$ which should provide help to theoreticians engaged in calculations of protein conformational stability. This is an area of ever increasing popularity which nevertheless suffers from a lack of credible potential energy functions likely to provide a true description of real interactions in an aqueous medium rather than in a vacuum or a crystal.

6. Towards Credible Structural and Dynamic Models of Alcohol–water Mixtures

Since the publication of the first review on alcohol–water mixtures the most significant experimental contributions to a better understanding of the behaviour of such systems have been the light scattering and SAXRS studies, the emphasis of thermodynamic measurements on very dilute solutions and the detailed analysis of the diffusion rates (rotational and translational) of the two molecular species in the mixtures. The *ab initio* calculations designed to provide credible water–alcohol and alcohol–alcohol pair potential functions have made possible computer simulations of infinitely and reasonably dilute solutions. Although such simulations have not produced any revolutionary new insights, the results fully support the conclusions reached earlier by a range of experimentalists. This lends credence to the pair potential functions on which the simulations are based.

A credible molecular model must be able to account for the observed behaviour of the mixtures over the whole range of temperature and pressure and to predict properties that have not yet been measured. Since much of the more recent experimental evidence is of thermodynamic origin, it is opportune to be reminded of the scope and limitations of such information, so concisely stated by Lewis and Randall: 'Thermodynamics exhibits no curiosity; certain things are poured into its hopper, certain other things emerge according to the laws of the machine, no cognizance being taken of the mechanism of the process or of the nature and character of the various molecular species concerned...In a thermodynamic formula we may use the value of a partial molal quantity equally well whether it is positive or negative, and if it proves to be the latter, no explanation need be given.' [**145**]

Since thermodynamics takes no account of the nature of molecules, any molecular model derived from thermodynamic measurements needs confirmation from non-thermodynamic sources, to make sure that the quantities and parameters used to fit the model to the experimental data are of a sensible magnitude. The converse also applies: any structural model based on results from scattering or spectroscopic measurements must be shown to be consistent with the balance of the various molecular interactions, as reflected in the thermodynamic properties of the mixtures in question.

A discussion of realistic models of alcohol–water mixtures is best attempted by considering separately several distinct concentration ranges. In a sense, such distinctions are somewhat artificial, but it is by now apparent that most physical properties change quite dramatically over very narrow ranges of concentration. Quite apart from that, however, rigorous molecular models are confined to solutions containing one or two solute molecules only. Beyond that, recourse must be made to chemical models.

6.1. *Infinite Dilution*

The infinitely dilute solution can be described in terms of two molecular pair distribution functions, the basic question being whether the observed physical properties result primarily from alcohol–water hydrogen bonding or from hydrophobic hydration, i.e. alkyl–water interactions. The magnitudes of the limiting partial molar quantities Y_2^0 reviewed in section 4 suggest that the latter effect predominates. The magnitudes of the Y_2^0 functions increase with increasing alkyl chain length, with V_2^0 and $C_{P,2}^0$ being linear functions of the number of CH_2 groups. A comparison of isomeric alcohols, e.g. the four butanols, shows that the alkyl group shape also contributes to the observed Y_2^0. Hydrophobic hydration involves a reorganization of water molecules to a quasi-clathrate type geometry which, according to the computer simulation data, extends to well beyond the primary hydration shell (see figure 24). Compared to the hydrophobic effect, alcohol–water hydrogen bonding provides only a small contribution to the observed behaviour, even in the case of MeOH. All the available evidence supports the picture of the alcohol as a 'soluble hydrocarbon' rather than an 'alkylated water'.

The temperature derivatives of the limiting partial molar properties are normal, in the sense that a rise in the temperature tends to decrease the extent of structural reorganization induced in water by the solute.

6.2. *The Dilute Solution: Solute Pair Interactions*

At finite concentrations solute–solute interactions give rise to the observed concentration dependence of measured physical properties. At the microscopic level the question arises whether alcohols interact primarily by direct hydrogen bonding or whether the observed effects are dominated by the interactions between 'hydrated' alkyl groups.

The postulate of two distinctive modes of hydration forms the basis of Pratt and Chandler's calculation of B^* of methanol in water. [**146**] They state that alkyl–alkyl correlations are affected by the polar –OH groups and coupled to them. This coupling is said to be responsible for the observation that $\mathrm{d}B^*/\mathrm{d}T$ is negative, with the authors dismissing earlier suggestions [**40, 109, 110**] that the hydrophobic interaction becomes more attractive at higher temperatures and that this is in fact responsible for negative $\mathrm{d}B^*/\mathrm{d}T$ values of *all* monofunctional alkyl derivatives. Pratt and Chandler [**147**] also state that, according to their theory of the hydrophobic effect [**147**], $\mathrm{d}B^*/\mathrm{d}T$ for non-polar solutes (e.g. methane) is positive. However, high precision vapour pressure measurements on aqueous solutions hydrocarbons have shown $\mathrm{d}B^*/\mathrm{d}T$ to be negative [**117**], suggesting that it is indeed hydrophobic effects which dominate the solution behaviour of simple monofunctional alkyl derivatives, at least in dilute solution. We therefore conclude that the microscopic theory of Pratt and Chandler does not correctly account for the solution behaviour of hydrocarbons and that it may be their choice of model for the methanol molecule which produces the correct sign for $\mathrm{d}B^*/\mathrm{d}T$, rather than any coupling between the effects of the alkyl group and the –OH group on solute interactions in solutions. This conclusion is also borne out by the computer studies on dilute solutions. [**140, 141**]

The complexity and anisotropy of the solute pair interaction has already been discussed and it is emphasized in figures 7 and 8 while g_{22} and h_{22} (for tert.-BuOH) have the expected signs, this is not the case for V_{22} and $C_{P,22}$, observations which led several authors to suggest that the alkyl–alkyl pair interaction might be of a solvent separated type, consistent with the crystalline analogue of the gas hydrate. Here again the computer studies on pairs of the Lennard–Jones spheres [**7**], methane molecules [**9**] and alcohol molecules [**140, 141**] are consistent with such earlier readings of experimental results.

Of particular relevance is the computed potential of mean force between two MeOH molecules shown in figure 25. [**140**] The authors state that not too much reliance can be placed on the detailed shape, but it is clear that the solvent separated state is preferred over the contact hydrophobic association [**142, 143, 144**], at least at 298 K. The shape of $W_{22}(r)$ is reminiscent of that calculated, but starting from a completely different set of premises, by Pratt and Chandler. [**146**] We would now speculate that a rise in temperature would change $W_{22}(r)$ such that the contact pair becomes more stable, at the expense of the water-separated pair stability. Such an effect would be consistent with the observed lower critical demixing phenomena in solutions of hydrophobic solutes. The above reasoning can account for the unexpected trends in the concentration dependence of $C_{P,2}$ and V_2. Once again the magnitudes of the anomalies increase with increasing alkyl chain length.

6.3. *The Moderately Dilute Solution, and Beyond*

The above model, based on molecular pair interactions only, fails completely at concentrations beyond those corresponding to the extrema in the $Y_2(x_2)$ relationships, as shown, for instance, in figure 9. However, even at substantially lower concentrations the higher virial coefficients in equations of the type (4.10) begin to contribute significantly to Y_2. In this concentration range the SAXRS and light scattering data are indicative of a rapidly increasing aggregation, almost akin to micelle formation by surfactants. The physical properties of mixtures in this concentration range are extremely sensitive to changes in temperature and pressure. The narrow maxima in the scattering intensity/concentration plots, see figure 6, have no counterpart in the thermodynamic functions. The latter suggest that once the extreme value has been exceeded, increasing concentration produces a more normal solution behaviour, water having lost its unique character and behaving almost like a polar organic liquid. The properties of such solutions are characterized by an almost zero temperature dependence.

Modern theoretical approaches based on statistical thermodynamics, as well as computer modelling techniques, can be usefully employed in studies of very dilute solutions where pair interactions predominate. They are not applicable to systems in which multiplet interactions are important, and especially where cooperative effects dominate the solution behaviour.

Reference has already been made to the Kirkwood–Buff theory of solutions the application of which permits the exact calculation of the integrals G_{ij} of the orientation averaged pair distribution functions; see section 3. It is thus possible to relate thermodynamic properties to solution 'structure'. This has been done for EtOH-water mixtures by Ben-Naim [26] and Donkersloot. [27] The results are summarized in figure 27. The G_{ij} functions are seen to be very

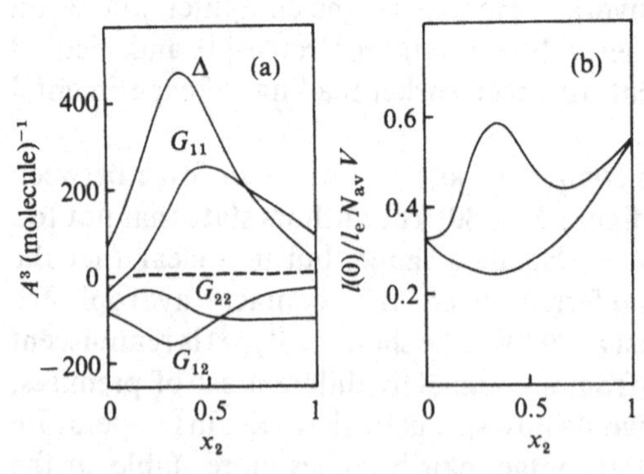

Figure 27. (a) Composition dependence of the Kirkwood–Buff integrals, as defined by eqns. (3.7), (3.9) and (3.10) for ethanol–water mixtures at 25 °C. The broken line indicates the compressibility contribution $kT\beta_T$, see eqn (3.8); (b) total zero angle X-ray scattering intensities (top curve) and compressibility contributions (bottom curve); I_e is the intensity scattered by a free electron, N_{av} is the average number of scattering species in volume V. After Donkersloot. [27]

sensitive to the composition of the mixture but at the present stage of our knowledge it is not possible to interpret $G_{ij}(x_2)$ in terms of molecular interactions. The details of the functions are probably governed mainly by the balance between the positive and negative contributions from the integrand $4\pi r^2\{g_{ij}(r)-1\}$ over the whole range of r. A significant feature of the calculations is the fact that the compressibility term hardly contributes to the G_{ij} functions but that the zero-angle scattering intensity $I(0)$ contains a large compressibility contribution. This suggests that wide angle and small angle scattering are related to different thermodynamic origins.

The data on which figure 27 is based are not the most up-to-date ones. More recent thermodynamic and scattering results, discussed in this review, should permit more accurate calculations of G_{ij} and $I(0)$ to be performed. These, in turn, will provide novel information about molecular interactions in concentrated solutions.

6.4. *Chemical Models*

The objective of chemical models is to provide fits to the observed properties of alcohol–water mixtures in terms of a (small) number of simultaneous equilibria. An early attempt by Mikhailov adopted as a starting point an interstitial model for water,[**149**] composed of a low density, ice-like species in equilibrium with a dense species of non-hydrogen bonded water the molecules of which could also occupy the cavities in the hydrogen bonded lattice. The solute, EtOH, competed with water for lattice sites but was partitioned between the lattice and the dense water with which it was assumed to mix and form a regular solution.

The following three equilibria were postulated:

(1) $(H_2O)_{\text{lattice}} \rightleftharpoons (H_2O)_{\text{interst.}}$
(2) $(EtOH)_{\text{interst.}} \rightleftharpoons (EtOH)_{\text{regular}}$
(3) $(H_2O)_{\text{interst.}} \rightleftharpoons (H_2O)_{\text{regular}}$

The three equilibrium constants between them yielded the varying proportions of the species with the EtOH concentration and temperature, and Mikhailov was able to fit $\Delta G_e\ (x_2)$, $V_2(x_2)$ and $H_2(x_2)$.

A clue to the structure of alcohol–water mixtures is provided by the dependence of the concentration trends on chain length of the alcohol. It was shown in section 4.2.7 that the rapid changes in all properties of alcohols become sharper and occur at lower concentrations the longer the alkyl chain length. In the case of n-alcohols the low solubility limits such comparison to the first three members, but n-BuOH can also be made miscible with water by the addition of small amounts of surfactants. Similarly the miscibility barrier can be overcome by increasing the hydrophilic character of the polar group. 2-butoxyethanol is completely miscible with water below 49 °C and its properties lie somewhere between the expected properties of n-BuOH and n-PenOH in water[**121**]. Longer alkyl chains can be investigated by the addition of extra ($-OCH_2CH_2-$) groups at the polar end or by the replacement of the OH group by a more hydrophilic group such as amine or phosphine oxides.

A mass-action model which can account quantitatively for the thermodynamic properties of non-ionic surfactants[**120**] has also been applied to alcohol–water mixtures up to medium concentrations. [**150**] The aggregation numbers obtained from the least-squares treatment (6 for tert.-BuOH and 4 for 2-PrOH) are of course not very realistic. Recently, Roux and Desnoyers have extended this mass-action model to cover the whole mol fraction range by assuming a double equilibrium, one involving the association of alcohols and one the association of water. [**151**] Both liquids in the pure state are assumed partially associated. The experimental thermodynamic data can then be fitted with a non-linear least-squares expression and the relevant parameters extracted. This is shown in figure 28 for the volume of mixing divided by the product of the two mol fractions of the n-PrOH-water mixtures. The fit is essentially quantitative with the minimum number of parameters which all have reasonable magnitudes: two changes in volume, two equilibrium constants related to $\Delta\mu_i$ and two aggregation numbers a_i. The fit is amazingly good considering that the structure of the hydrophobic cosphere is assumed identical to the structured part of liquid water, the size of the clusters of

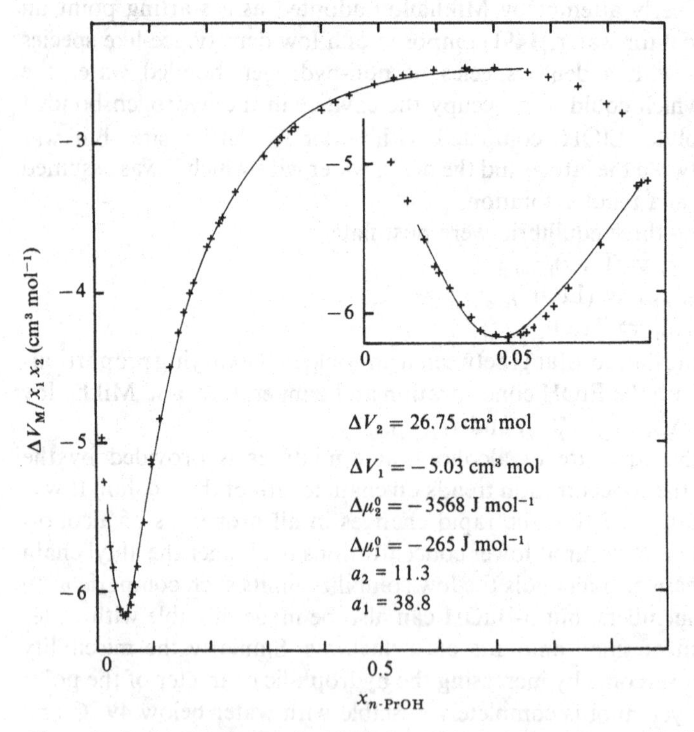

Figure 28. Reduced volumes of mixing of n-PrOH and water. Data points from reference 54 and full line from the double equilibria model of Roux and Desnoyers. [**151**]

water and of alcohols are assumed independent of concentration, specific hydrogen-bonding interactions between alcohols and water are ignored, etc. Therefore no physical significance can be attached to the numbers that are needed to generate the shape of the thermodynamic quantities. In fact any double equilibrium can in principle be made to fit such trends. For example, Hvidt and coworkers have succeeded in fitting the volumes of alcohol–water mixtures, using a model based on the dehydration of the alkyl chain and of the OH group. [**52**] The success is surprising since they assume the wrong sign for the volume contribution to hydrophobic hydration ($\Delta V < 0$) and it is expected from the discussion on the properties of water in alcohols (section 4), that the alcohols would not be dehydrated until very high concentrations are reached, whereas the model suggests that at $x_2 \simeq 0.1$ the alkyl groups are almost completely unsolvated.

Another promising double equilibrium model for alcohol–water mixtures is probably that of Fujiyama *et al.* [**34–36, 152, 153**] They were able to account for the trends in the concentration dependence of the Rayleigh scattering, reduced to mutual diffusion coefficients, by the following set of equilibria.

$$A + iH_2O \rightleftharpoons A(H_2O)_i$$
$$jA(H_2O)_i \rightleftharpoons \{A(H_2O)_i\}_j$$
$$\{A(H_2O)_i\}_j \rightleftharpoons jA + ijH_2O.$$

Figure 29 illustrates this model for tert.-BuOH in a pictorial fashion. For $x_2 < 0.5$ the species $A(H_2O)_{20}$ predominates, while the equilibria involving $\{A(H_2O)_{20}\}_4$ mostly account for the data above $x_2 \simeq 0.05$. This model is appealing since it does support the clathrate hydrate nature of hydrophobic hydration and explains in a simple fashion why many of the partial molar properties of alcohols at $x_2 > 0.5$ are so similar to the pure liquid alcohol values. It also accounts for the success of the micelle models for alcohol–water mixtures since the local regions where free alcohol molecules predominate are not unlike small micellar aggregates. Nevertheless, like most of the chemical models, this approach oversimplifies the structure of the alcohol–water

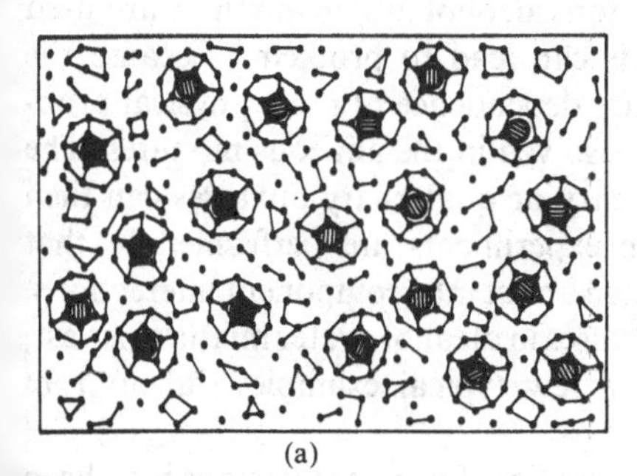

(a)

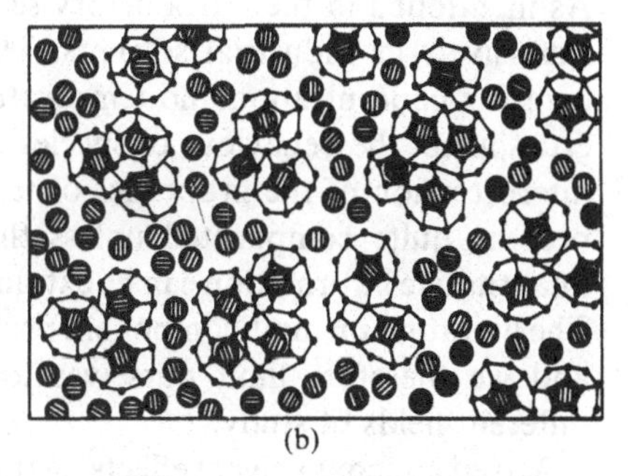

(b)

Figure 29. Tert.-BuOH–water model of Iwasaki and Fujiyama: (a) $x_2 \simeq 0.02$; (b) $x_2 \simeq 0.1$. Reproduced, with permission, from reference 35.

mixtures. In particular, it fails to take into account the highly developed structure of liquid water, the formation of solvent-shared solute pairs at low concentration where fewer water molecules are involved (an intermediate species of the type $\{A(H_2O)_i\}_2$ where $i < 20$ should exist), and the extensive hydrogen bonding known to exist between alcohols and water as $x_2 > 0.5$. Most important, however, is the fact that the model is incompatible with statistical thermodynamics, since it makes no provision for the limiting case of pair interactions, expressed through B^*, the osmotic second virial coefficient. This is of course a stricture which can be applied to all chemical equilibrium models. If all the above effects are taken into account, then either the problem becomes too complex mathematically or it requires so many adjustable parameters that it could be made to fit the proverbial elephant. A successful chemical model for alcohol–water mixtures will have to be consistent with the well-established three-dimensional nature of water, the clathrate-like nature of hydrophobic hydration, the pairwise nature of hydrophobic interactions at low concentration and the predominance of extensive hydrogen bonding between water and alcohols up to very high mol fractions of alcohols and yet be sufficiently simple to be handled mathematically. None of the existing models meets all these conditions.

Only for the sake of completeness do we make reference to attempts to fit (force fit is a more appropriate expression) the physical properties of alcohol–water mixtures by polynomial expansions without regard to 'chemical common sense'. The mixture is thus treated as a fully homogeneous system in which solute–water gradually replace water–water interactions, to be replaced in turn by solute–solute interactions as $x_2 \rightarrow 1$. In view of the rather unorthodox concentration dependences of nearly all measured physical properties, this type of polynomial fitting requires a quite unrealistically large number of adjustable parameters, up to 12 in some cases, none of which has any physical significance.[**54**]

7. Aqueous Alcohol Solutions as Solvent Media

As mentioned in the introductory section, alcohol–water mixtures are used extensively as industrial solvents. This can lead to problems, because the complex concentration and temperature dependence of most physical properties affect the behaviour of species dissolved in the aqueous mixtures. The interpretation of the properties of ternary or quaternary mixtures can then become quite complicated unless the experiments are performed so that quantitative information can be extracted about intercomponent interactions. The literature on multicomponent processes in alcohol–water mixtures is vast, and we therefore limit ourselves to a few typical examples, taken from different fields of study.

Mixed aqueous solvent effects on the kinetics of organic reactions have been recently reviewed.[**154**] Mixed solvent effects on solvolytic reactions have been particularly well investigated. As a typical example we chose the neutral pH-independent hydrolysis of p-nitrophenyl dichloroacetate (I)

$HCCl_2.COO.C_6H_4.NO_2$–p

(I)

in tert.-BuOH/H_2O and tert.–BuOD/D_2O mixtures.[155] The activation parameters, shown in figure 30 as a function of the solvent composition, are typical. No simple correlation exists between $\Delta G^{\ddagger}$ and the dielectric permittivity of the solvent. In order to investigate the solvent effects on the ground state and the transition state, the authors chose p-nitrophenyl acetate (II) as a model of the transition state

$CH_3.COO.C_6H_4.NO_2$–p

(II)

and measured the thermodynamic transfer functions of I and II for the process:

I,II (H_2O or D_2O)$\rightarrow$I,II (mixed solvent of composition x_2).

The rate effect in figure 30 was found to be due to differences in solvation of the ground state in the sense that tert.-BuOH(D) stabilizes the reactant species I. Dramatic changes take place in $\Delta H^{\ddagger}$ and $T\Delta S^{\ddagger}$ with increasing x_2. At $x_2 = 0$ the contributions of $\Delta H^{\ddagger}$ and $T\Delta S^{\ddagger}$ towards $\Delta G^{\ddagger}$ differ by only 7.5 kJ mol^{-1}, but at $x_2 = 0.05$ this difference amounts to approximately 40 kJ mol^{-1}. The rate determining step in the reaction is said to involve a proton transfer (from the solvent) and extensive solvent immobilization in the transition state. Clearly the hydrophobic hydration (which is maximized at

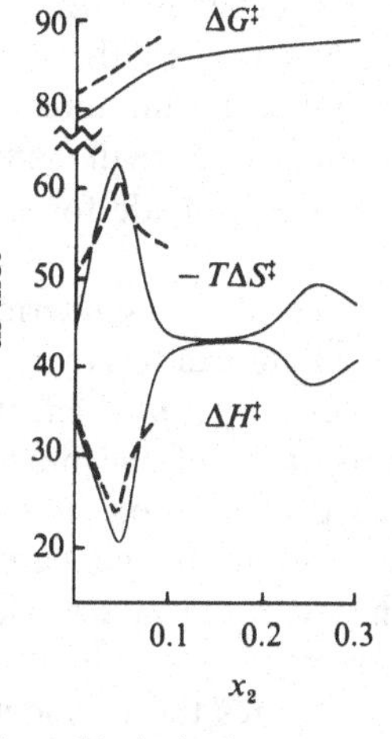

Figure 30. Activation parameters *vs.* x_2 for the neutral hydrolysis of *I* in tert.-BuOH–H_2O (solid line) and tert.-BuOD–D_2O (broken line) mixtures at 25 °C.[155]

$x_2 = 0.05$) affects this process but, as usual, enthalpy–entropy compensation results in a reasonably well-behaved $\Delta G^{\ddagger}$ (x_2) profile. The most pronounced mixed solvent effects are reflected in $\Delta C_p^{\ddagger}(x_2)$. Thus, for the solvolysis of tert.-BuCl in tert.-BuOH–water mixtures $\Delta C_p^{\ddagger}$ (J mol^{-1} K^{-1}) changes from -330 at $x_2 = 0$, to -794 at $x_2 = 0.05$ and -50 at $x_2 = 0.10$, to level off at -63 for $x_2 > 0.2$. [**156**]

An almost identical heat capacity-solvent composition relationship exists for the unfolding of ribonuclease in EtOH-water mixtures. [**157**] This gives rise to a complex ΔG (x_2, T) profile. At low temperatures (< 30 °C) and $0 < x_2 < 0.1$, EtOH *stabilizes* the native enzyme, whereas at temperatures $\geqslant$ 20 °C the enzyme is destabilized for all values of x_2, compared to its stability in aqueous solution.

At first sight such ΔG (x_2, T) behaviour might be rationalized in terms of preferential water or EtOH binding processes, but the futility of such an approach becomes obvious from a comparison of the above results with the effects of EtOH–water mixtures on the solubility of argon. An almost identical $\Delta G_{\mathrm{solution}}$ (x_2, T) is observed, with argon 'salted-out' of solution at low temperatures and low EtOH concentrations, with the expected solubility enhancement taking over at approximately 20 °C for all values of x_2. [**158**] It becomes apparent that the monatomic rare gas and the complex native protein respond in an identical manner to the mixed aqueous solvent.

A rigorous thermodynamic analysis of alcohol–water–salt mixtures has been performed by Desnoyers *et al.* [**159**] taking into account the microheterogeneity of the solvent mixture. Figure 31 illustrates the thermodynamic functions of transfer of NaCl to tert.-BuOH–water mixtures. At low concentration the concentration dependences of the transfer of NaCl to tert.-BuOH solutions are identical to the transfer of tert.-BuOH to NaCl solutions, as expected from the reciprocity theorem. The initial slope reflects the pair interaction between NaCl and tert.-BuOH. At higher tert.-BuOH concentrations the trends in the transfer functions of NaCl reflect the shift in the equilibrium between different states of the alcohol–water mixtures in the presence of NaCl. If the electrolyte is hydrophobic (a large tetraalkylammonium halide or a surfactant) then allowance must also be made for the formation of mixed micelles or mixed aggregates.

Attempts have also been made to rationalize the transfer functions in terms of the scaled particle theory, according to which the solute has to create a cavity in the solvent, no explicit account being taken of 'structural' effects in the solvent. [**160**] However, since the physical properties of the mixed solvent which themselves reflect such structural effects are used in the calculations, it is not too surprising that the signs and trends in the calculated transfer functions correspond reasonably well with the measured values. The agreement in the magnitudes is hardly acceptable, however.

Despite their limitations for binary alcohol–water mixtures those models involving microphase transitions can account for the properties of various

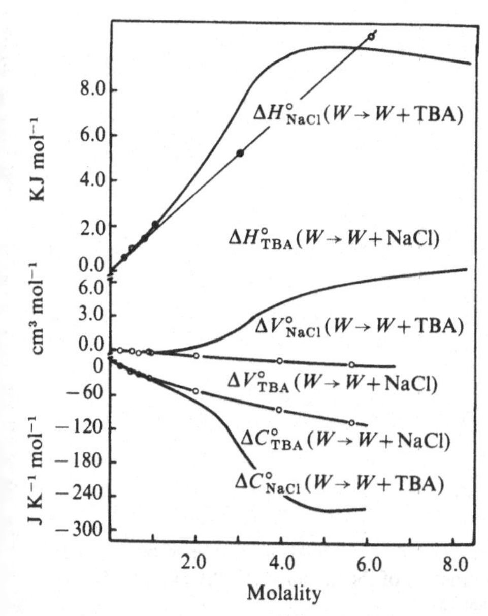

Figure 31. Standard thermodynamic functions of transfer of the system tert.-BuOH–NaCl–water at 25 °C. Reproduced, with permission, from reference 159.

solutes dissolved in alcohol–water mixtures, and in some cases the models are nearly quantitative. Recently a chemical equilibrium model was proposed for mixed micellar systems. [**150**] Essentially the thermodynamic transfer functions of a hydrophobic solute from water to a micellar system are related to three effects:

(a) Pair interactions between the solute and the monomers of the micellar system. These predominate in the pre-micellar region.

(b) Distribution of the solute between water and the micelle. The molar properties of the solute in the micelle are assumed to be similar to those of the pure liquid solute.

(c) Shifts in the monomer–micelle equilibrium caused by the presence of the solute in the micelle. This effect is responsible for the observed extrema in the transfer functions.

Using a simple mass-action model for the micellization process and estimating the distribution constant of the solute from its solubility in water, it is possible to explain the observed trends in the transfer functions by the use of data from the binary systems only. This is illustrated in figure 32 for the system 2-PrOH–water–benzene. [**161**] For the calculations the alcohol is assumed to form small micelles and the mass-action model yields an aggregation number of 4. The pair interaction term was neglected and the only parameter adjusted was the distribution constant of benzene between water and the alcohol 'micelles'. The fit is seen to be nearly quantitative.

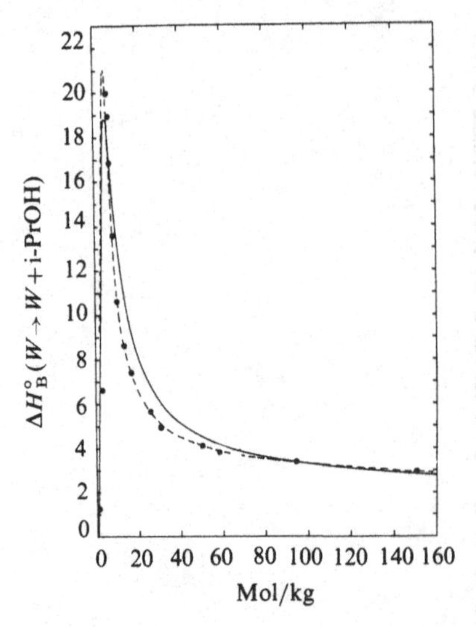

Figure 32. Simulated (full time) and experimental (●) enthalpies of transfer of benzene (B) from water to i-PrOH–water mixtures as a function of the molality of i-PrOH at 25 °C. Reproduced, with permission, from reference 150.

Simple phase or chemical equilibrium treatments of solutes in alcohol–water mixtures are not limited to thermodynamic properties. For example, Palma and Morel were able to rationalize the viscosities of alkali halides and of urea in tert.-BuOH solutions, using a similar approach. [**162, 163**] Presumably other properties, such as conductance, could equally well be explained without any need to invoke changes in the structure of the solvent medium other than those resulting from the relative concentrations of the various species of alcohol–water complexes.

Finally, it is important to note that many complex systems, such as microemulsions, depend significantly on the peculiar properties of alcohols and other similar organic cosolvents in water. Microemulsions are thermodynamically stable mixtures of oil, water, surfactant and cosurfactants (usually medium chain length alcohols). These systems are used in industry, medicine, oil recovery, etc. The role of the cosurfactant or alcohol has been underestimated until relatively recently. However, as shown in figure 33, the thermodynamic properties of ternary systems like water–benzene–2-PrOH [**161**] are remarkably similar to those of microemulsions containing surfactants and cosurfactants as active mixtures. [**164, 165**] This suggests that the structure of these microemulsion may not be very different from the microheterogeneities that already exist in binary water–alcohol mixtures.

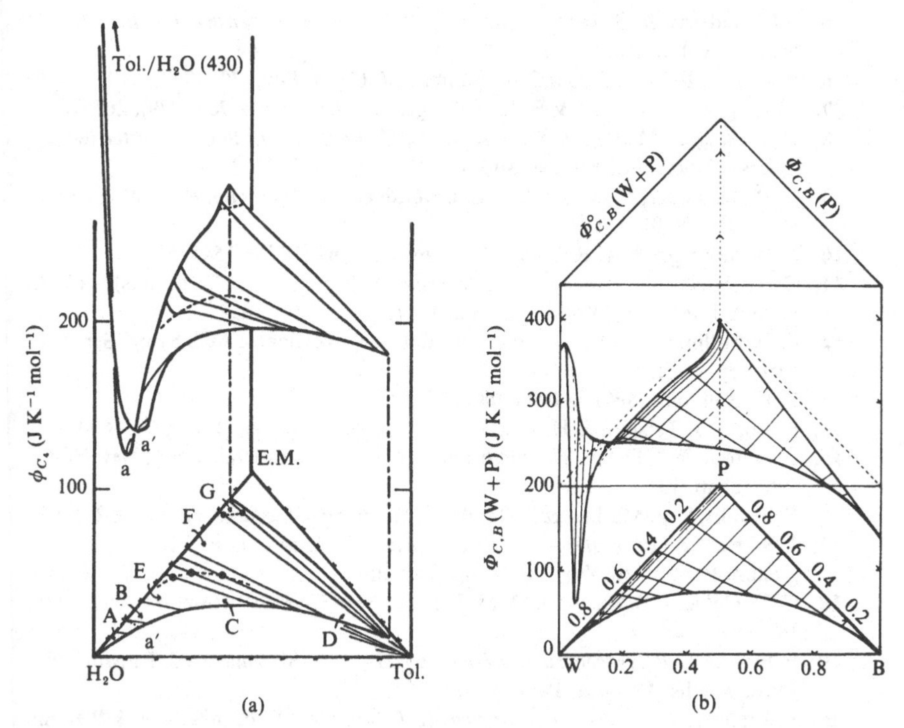

Figure 33. Apparent molar heat capacities of toluene or benzene in ternary or pseudo-ternary systems at 25 °C: (a) toluene in the system water–toluene–sodium dodecylsulfate-n-BuOH for a fixed ratio of surfactant to alcohol of 1:2 (EM); (b) benzene in the system water–benzene–2-PrOH. Reproduced, with permission, from references 161 and 164.

8. Conclusion

The previous review[**1**] concluded with a quotation from Scatchard[**166**]: 'The best advice which comes from years of study of liquid mixtures is to use any model in so far as it helps, but not to believe that any moderately simple model corresponds very closely to any real mixture.'

This was timely advice in 1966, but we would now venture to claim that our understanding of the real nature of alcohol–water mixtures is on a much firmer basis.

References

1. F. Franks & D. J. G. Ives. *Quart. Rev. Chem. Soc.* **20** (1966), 1–44.
2. F. Franks. *Faraday Symp. Royal. Soc. Chem.* **17** (1982), 7–10.
3. D. Wood. In *Water – A Comprehensive Treatise* (ed. F. Franks), vol. 6. Plenum Press: New York, 1979, pp. 279–409.
4. A. H. Narten & H. A. Levy. In *Water – A Comprehensive Treatise* (ed. F. Franks), vol. 1. Plenum Press: New York, 1972, pp. 311–32.

5. J. E. Enderby & G. W. Nielson. In *Water – A Comprehensive Treatise* (ed. F. Franks), vol. 6. Plenum Press: New York, 1979, pp. 1–46.
6. A. Geiger, F. H. Stillinger & A. Rahman. *J. Chem. Phys.* **70** (1979), 4185–4193.
7. A. Geiger, A. Rahman & F. H. Stillinger. *J. Chem. Phys.* **70** (1979), 263–76.
8. C. S. Pangali, M. Rao & B. J. Berne. In *Computer Modeling of Matter* (ed. P. Lykos). American Chemical Society: Washington, D.C., 1978.
9. G. Ravishanker, M. Mezei & D. L. Beveridge. *Faraday Symp. Royal Soc. Chem.* **17** (1982), 79–91.
10. F. H. Stillinger & A. Rahman. *J. Chem. Phys.* **60** (1974), 1545–57.
11. C. A. Angell. In *Water – A Comprehensive Treatise* (ed. F. Franks), vol. 7. Plenum Press: New York, 1982, pp. 1–81.
12. F. H. Stillinger. In *Water in Polymers* (ed. S. P. Rowland), ACS Symp. Ser. 127, 1980, pp. 11–21.
13. F. H. Stillinger. *Science* **209** (1980), 451–7.
14. L. Bosio, J. Teixeira & H. E. Stanley. *Phys. Rev. Lett.* **46** (1981), 597–600.
15. L. Dufour & R. Defay. *Thermodynamics of Clouds*. Academic Press: New York, 1963, chap. 13.
16. F. Franks, S. F. Mathias & K. Trafford. *Colloids and Surfaces*, **11** (1984), 275–85.
17. J. B. Hasted. In *Water – A Comprehensive Treatise* (ed. F. Franks), vol. 1. Plenum Press: New York, 1972, pp. 255–309.
18. C. H. Collie, J. B. Hasted & D. M. Ritson. *Proc. Phys. Soc.* (London), **60** (1948), 145–60.
19. F. Franks. In *Polysaccharides in Foods* (eds. J. M. V. Blanshard & J. R. Mitchell). Butterworths: London, 1979, pp. 33–50.
20. F. Franks. In *Water – A Comprehensive Treatise* (ed. F. Franks), vol. 4. Plenum Press: New York, 1975, pp. 1–93.
21. S. Okazaki, K. Nakanishi & H. Touhara. *J. Chem. Phys.* **78** (1983), 454–69.
22. K. Nakanishi, K. Ikari, S. Okazako & H. Touhara. *J. Chem. Phys.* **80** (1984), 1656–70.
23. P. S. Ramanathan & H. L. Friedman. *J. Chem. Phys.* **54** (1971), 1086–99.
24. H. L. Friedman & C. V. Krishnan. *J. Solution Chem.* **2** (1973), 119–38.
25. J. G. Kirkwood & F. P. Buff. *J. Chem. Phys.* **19** (1951), 774–7.
26. A. Ben-Naim. *Water and Aqueous Solutions. Introduction to a Molecular Theory*. Plenum: New York, 1974, chap. 4; also *J. Chem. Phys.* **67** (1977), 4884–90.
27. M. C. A. Donkersloot. *J. Solution Chem.* **8** (1979), 293–307.
28. K. Kojima, T. Kato & H. Nomura. *J. Solution Chem.* **13** (1984), 151–65.
29. J. J. Kozak, W. S. Knight & W. Kauzmann. *J. Chem. Phys.* **48** (1968), 675–90.
30. E. G. Richards. *An Introduction to Physical Properties of Large Molecules in Solution*. Cambridge University Press, 1980.
31. E. A. Guggenheim. *Mixtures*. Oxford University Press, 1952.
32. R. A. Robinson & R. H. Stokes. *J. Phys. Chem.* **65** (1961), 1954–8.
33. F. Cennamo & E. Tartaglione. *Nuovo Cimento* **11** (1959), 401–9.
34. K. Iwasaki, M. Tanaka & T. Fujiyama. *Bull. Chem. Soc. Japan* **49** (1976), 2719–23.
35. K. Iwasaki & T. Fujiyama. *J. Phys. Chem.* **81** (1977), 1908–12.
36. K. Iwasaki & T. Fujiyama. *J. Phys. Chem.* **83** (1979), 463–8.
37. H. D. Bale, R. E. Shepler & D. K. Sorgen. *Phys. Chem. Liquids* **1** (1968), 181–90.
38. F. Franks, M. J. Quickenden, D. S. Reid & B. Watson. *Trans. Faraday Soc.* **66** (1970), 582–9.

39. S. Cabani, P. Gianni, V. Mollica & L. Lepori. *J. Solution Chem.* **10** (1981), 563–95.
40. S. Westmeier. *Chem. Tech.* (Leipzig) **28** (1976), 350–3; **28** (1976), 480–3; **29** (1977), 218–22; **30** (1978), 354–7.
41. H. Mitsutake & M. Sakai. *J. Chem. Eng. Japan* **8** (1975), 435–9.
42. R. C. Pemberton & C. J. Mash. *J. Chem. Thermodyn.* **10** (1978), 867–88.
43. R. O. Pryanikova & G. D. Efremova. *Fiz. Khim. Rastvarov* (1972), 228.
44. J. A. Larkin. *J. Chem. Thermodyn.* **7** (1975), 137–48.
45. B. Y. Okamoto, R. H. Wood & P. T. Thompson. *J. Chem. Soc. Faraday Trans.* 1 **74** (1978), 1990–2007.
46. S. D. Christian, E. H. Lane & E. E. Tucker. *J. Solution Chem.* **10** (1981), 181–8.
47. F. Franks, M. Pedley & D. S. Reid. *J. Chem. Soc. Faraday Trans.* 1 **72** (1976), 359–67.
48. J. E. Desnoyers, G. Perron, L. Avedikian & J.-P. Morel. *J. Solution Chem.* **5** (1976), 631–44.
49. G. Perron & J. E. Desnoyers. *J. Chem. Thermodyn.* **13** (1981), 1105–21.
50. C. Jolicoeur. Thermodynamic flow methods in biochemistry. Calorimetry, Densimetry and dilatometry. *Methods of Bioch. Anal.* **27** (1981), 171–287.
51. J. F. Alary, M. A. Simard, J. Dumont & C. Jolicoeur. *J. Solution Chem.* **11** (1982), 755–76.
52. C. Dethlefsen, P. G. Sørensen & A. Hvidt. *J. Solution Chem.* **13** (1984), 191–202.
53. K. Nakanishi, N. Kato & M. Maruyama. *J. Phys. Chem.* **71** (1976), 814–8.
54. C. G. Benson & O. Kiyohara. *J. Solution Chem.* **9** (1980), 791–804.
55. C. DeVisser, G. Perron & J. E. Desnoyers. *Can. J. Chem.* **55** (1977), 856–62.
56. G. Roux, D. Roberts, G. Perron & J. E. Desnoyers. *J. Solution Chem.* **9** (1980), 629–47.
57. L. Bøje & A. Hvidt. *J. Chem. Thermodyn.* **3** (1971), 663.
58. S. G. Bruun & A. Hvidt. *Ber. Bunsenges. Phys. Chem.* **81** (1977), 930–3.
59. A. Hvidt, R. Moss & G. Nielson. *Acta Chem. Scand.* **B32** (1978), 274–80.
60. K. N. Marsh & A. E. Richards. *Austral. J. Chem.* **33** (1980), 2121–32.
61. J. P. Grolier & E. Wilhelm. *Fluid Phase Equil.* **6** (1981), 283–7.
62. N. A. Agaev, A. A. Pashaev & A. M. Kerimov. *Zhur. Fiz. Khim.* **49** (1975), 3021; **48** (1974), 1616.
63. U. N. Popov & B. A. Malov. *Zhur. Fiz. Khim.* **45** (1971), 2944; see also *Tr. Mosk. Energ. Inst.* **111** (1972), 129.
64. F. Franks & D. S. Reid. In *Water – A Comprehensive Treatise* (ed. F. Franks), vol. 2. Plenum Press: New York, 1973, pp. 323–80.
65. F. Franks & H. T. Smith. *Trans. Faraday Soc.* **64** (1968) 2962–72.
66. H. Høiland. *J. Solution Chem.* **9** (1980) 857–66.
67. S. Cabani, G. Conti & E. Matteoli. *J. Solution Chem.* **5** (1976) 751–63.
68. S. Cabani, G. Conti & L. Lepori. *J. Phys. Chem.* **78** (1974), 1030–4.
69. T. Nakajima, T. Komatsu & T. Nakagawa. *Bull. Chem. Soc. Japan* **48** (1975), 783–7.
70. S. Harada, T. Kakajima, T. Komatsu & T. Nakagawa. *J. Solution Chem.* **7** (1978), 463–74.
71. F. Franks. *J. Chem. Soc. Faraday Trans.* 1 **73** (1977), 830–2.
72. H. Høiland & E. Vikingstad. *Acta Chem. Scand.* *A***30** (1976), 182–6.
73. C. Jolicoeur & G. Lacroix. *Can. J. Chem.* **54** (1976), 624–31.
74. M. Manabe & M. Koda. *Bull. Chem. Soc. Japan* **48** (1975), 2367–71.

75. M. Sakurai & T. Nakagawa. *J. Chem. Thermodyn.* **16** (1984), 171–4.
76. A. D'Aprano & V. Agrigento. *Gazz. Chim. Ital.* **108** (1978), 703–4.
77. O. Kiyohara & C. G. Benson. *J. Solution Chem.* **10** (1981), 281.
78. J. Lara & J. E. Desnoyers. *J. Solution Chem.* **10** (1981), 465.
79. N. M. Murthy & S. V. Subrahmanyam. *Acoustica* **40** (1978), 263–8.
80. M. Nakagawa, H. Inubushi & T. Moriyoshi. *J. Chem. Thermodyn.* **13** (1981), 171.
81. H. Høiland & E. Vikingstad. *Acta Chem. Scand.* *A***30** (1976), 692–6.
82. T. Nakajima, T. Komatsu & T. Nakagawa. *Bull. Chem. Soc. Japan* **48** (1975), 788–90.
83. S. Harada, T. Nakajima, T. Komatsu & T. Nakagawa. *J. Solution Chem.* **7** (1978), 463–74.
84. S. Cabani, G. Conti & E. Matteoli. *J. Solution Chem.* **8** (1979), 11–23.
85. G. C. Benson, P. J. D'Arcy & O. Kiyohara. *J. Solution Chem.* **9** (1980), 931–8.
86. M. A. Anisimov, V. S. Esipov, V. M. Zaprudskii, N. S. Zaugol'nikova, G. I. Ovodov, T. M. Dvodova & A. L. Seifer. *J. Struct. Chem.* **18** (1977), 663–70.
87. G. C. Benson & P. J. D'Arcy. *J. Chem. Eng. Data* **27** (1982), 439–42.
88. G. S. Arutyunyan & A. D. Stepanakert. *Dokl. Akad. Nauk. SSSR* **274** (1984), 833–6.
89. E. M. Arnett, W. B. Kover & J. V. Carter. *J. Amer. Chem. Soc.* **91** (1969), 4028–34.
90. N. Nichols, R. Sköld, C. Spink, J. Suurkuusk & I. Wadsö. *J. Chem. Thermodyn.* **8** (1976), 1081–93.
91. J. P. Guthrie. *Can. J. Chem.* **55** (1977), 3700–06.
92. G. Perron & J. E. Desnoyers. *Fluid Phase Equil.* **2** (1979), 239–62.
93. P. M. Kessel'man, A. K. Voronyuk' & I. V. Onyfriev. *J. Appl. Chem.* **48** (1975), 2244–7.
94. A. V. Shal'gin & L. V. Puchkov. *Zh. Obshch. Khim.* **53** (1983), 2181–4.
95. C. M. Sorensen. *J. Chem. Phys.* **79** (1983), 1455–61.
96. M. Nakagawa, H. Inubushi & T. Moriyoshi. *J. Chem. Thermodyn.* **13** (1980), 171.
97. T. Moriyoshi, Y. Aoki & H. Kamiyama. *J. Chem. Thermodyn.* **9** (1977), 495–502.
98. Y. Aoki & T. Moriyoshi. *J. Chem. Thermodyn.* **10** (1978), 1173–9.
99. T. Moriyoshi & Y. Aoki. *J. Chem. Eng. Japan* **11** (1978), 341–5.
100. J. B. Ott, J. R. Goates & B. A. Waite. *J. Chem. Thermodyn.* **11** (1979), 739–46.
101. G. Wada & S. Umeda. *Bull. Chem. Soc. Japan* **35** (1962), 646–52; **35** (1962), 1797–1801.
102. F. Franks & B. Watson. *Trans. Faraday Soc.* **63** (1967), 329–34.
103. J. L. Neal & D. A. I. Goring. *J. Phys. Chem.* **74** (1970), 658–64.
104. D. D. MacDonald & J. B. Hyne. *Can. J. Chem.* **54** (1976), 3073–6.
105. S. V. Subrahmaniam & N. M. Murthy. *J. Solution Chem.* **4** (1975), 347–58.
106. D. D. MacDonald & J. B. Hyne. *Can. J. Chem.* **49** (1971), 2636–42.
107. D. D. MacDonald. *Can. J. Chem.* **54** (1976), 3559–63.
108. R. C. Wilhoit & B. J. Zwolinski. *J. Phys. Chem. Ref. Data* **2** Suppl. 1 (1973).
109. J. J. Savage & R. H. Wood. *J. Solution Chem.* **5** (1976), 733–50.
110. A. H. Clark, F. Franks, M. Pedley & D. S. Reid. *J. Chem. Soc. Faraday Trans.* 1 **73** (1977), 290–305.
111. F. Franks & M. D. Pedley. *J. Chem. Soc. Faraday Trans.* 1 **79** (1983), 2249–60.

112. A. Ben-Naim. *Hydrophobic Interactions*. Plenum Press: New York, 1980.
113. I. R. Tasker & R. H. Wood. *J. Phys. Chem.* **86** (1982), 4040–5.
114. K. Shinoda & M. Fujihara. *Bull. Chem. Soc. Japan* **41** (1968), 2612–15.
115. D. Patterson & M. Barbe. *J. Phys. Chem.* **80** (1976), 2435–6.
116. R. Lumry, E. Battiste & C. Jolicoeur. *Faraday Symp. Royal Soc. Chem.* **17** (1982), 93–108.
117. E. E. Tucker & S. D. Christian. *J. Phys. Chem.* **83** (1979), 426–7.
118. G. Barone, G. Castronuovo & V. Elia. In preparation.
119. J. E. Desnoyers, G. Caron, R. Delisi, D. Roberts, A. Roux & G. Perron. *J. Phys. Chem.* **87** (1983), 1397–1406.
120. G. Roux-Desgranges, A. H. Roux, J. P. Grolier & A. Viallard. *J. Solution Chem.* **11** (1982), 357–75.
121. G. Roux, G. Perron & J. E. Desnoyers. *J. Solution Chem.* **7** (1978), 639–54.
122. Y. DeGrandpre, J. B. Rosenholm, L. L. Lemelin & C. Jolicoeur. In *Solution Behavior of Surfactants*, Vol. 1 (eds. K. L. Mittal & E. J. Fendler). Plenum Press: New York, 1981, pp. 431–53.
123. E. v. Goldammer & H. G. Hertz. *J. Phys. Chem.* **74** (1970), 3734–55.
124. E. v. Goldammer & M. D. Zeidler. *Ber. Bunsenges. Phys. Chem.* **73** (1969), 4–15.
125. M. D. Zeidler. In *Water – A Comprehensive Treatise* (ed. F. Franks), vol. 2. Plenum Press: New York, 1973, pp. 529–73.
126. H. Leiter, K. J. Patil & H. G. Hertz. *J. Solution Chem.* **12** (1983), 503–17.
127. F. Franks, J. R. Ravenhill, P. A. Egelstaff & D. I. Page. *Proc. Roy. Soc.* **A319** (1970), 189–208.
128. T. T. Herskovits & T. M. Kelly. *J. Phys. Chem.* **77** (1973), 381–8.
129. G. Jones & M. Dole. *J. Amer. Chem. Soc.* **51** (1929), 2950–64.
130. R. L. Kay. In *Water – A Comprehensive Treatise* (ed. F. Franks), vol. 3. Plenum Press: New York, 1973, pp. 173–209.
131. A. Holtzer & M. F. Emerson. *J. Phys. Chem.* **73** (1969), 26–33.
132. R. L. Kay, unpublished data, cited by F. Franks in *Physicochemical Processes in Mixed Aqueous Solvents* (ed. F. Franks). Heinemann: London, 1966, p. 50.
133. M. J. Blandamer. In *Water – A Comprehensive Treatise* (ed. F. Franks), vol. 2. Plenum Press: New York, 1973, pp. 459–94.
134. M. J. Blandamer. In *Water – A Comprehensive Treatise* (ed. F. Franks), vol.2. Plenum Press: New York, 1973, pp. 495–528.
135. K. W. Miller, N. V. Reo, A. J. M. Schoot Uiterkamp, D. P. Stengle, T. R. Stengle & K. L. Williamson. *Proc. Nat. Acad. Sci. U.S.A.* **78** (1981), 4946–9.
136. J. A. Ripmeester & D. W. Davidson. *J. Mol. Struct.* **75** (1981), 67–72.
137. T. R. Stengle, S. M. Hosseini, H. G. Basiri & K. L. Williamson, *J. Solution Chem.* **13** (1984), 779–87.
138. J. E. Del Bene. *J. Chem. Phys.* **55** (1971), 4633–6.
139. G. Alagona & A. Tani. *J. Chem. Phys.* **74** (1981), 3980–8.
140. S. Okazaki, H. Touhara & K. Nakanishi. *J. Chem. Phys.* **81** (1984), 890–4.
141. H. Tanaka, K. Nakanishi & H. Touhara. *J. Chem. Phys.* **81** (1984), 4065–73.
142. W. Kauzmann. *Adv. Protein Chem.* **14** (1959), 1–63.
143. G. Nemethy & H. A. Scheraga. *J. Phys. Chem.* **66** (1962), 1773–89.
144. C. Tanford. *The Hydrophobic Effect*. Wiley & Sons: New York, 2nd edn, 1980.
145. G. N. Lewis & M. Randall. *Thermodynamics*. McGraw-Hill: New York, 2nd edn, 1961, p. 378.

146. L. R. Pratt & D. Chandler. *J. Solution Chem.* **9** (1980), 1–18.
147. L. R. Pratt & D. Chandler. *J. Chem. Phys.* **67** (1977), 3683–704.
148. V. A. Mikhailov. *J. Struct. Chem.* **9** (1968), 332–9.
149. H. S. Frank & A. Quist. *J. Chem. Phys.* **34** (1961), 604–11.
150. A. H. Roux, D. Hetu, G. Perron & J. E. Desnoyers. *J. Solution Chem.* **13** (1984), 1–25.
151. A. H. Roux & J. E. Desnoyers. In preparation.
152. N. Ito, I. Kato & T. Fujiyama. *Bull. Chem. Soc. Japan* **54** (1981), 2573–8.
153. N. Ito, K. Saito, T. Kato & T. Fujiyama. *Bull. Chem. Soc. Japan* **54** (1981), 991–7.
154. J. B. F. N. Engberts. In *Water – A Comprehensive Treatise* (ed. F. Franks), vol. 6. Plenum Press: New York, 1979, pp. 139–237.
155. J. F. J. Engberson & J. B. F. N. Engberts. *J. Amer. Chem. Soc.* **97** (1975), 1563–8.
156. R. E. Robertson & S. E. Sugamori. *Can. J. Chem.* **50** (1972), 1353–60.
157. J. F. Brandts & L. Hunt. *J. Amer. Chem. Soc.* **89** (1967), 4826–38.
158. A. Ben-Naim & S. Baer. *Trans. Faraday Soc.* **60** (1964), 1736–41.
159. J. E. Desnoyers, G. Persson, J. P. Morel & L. Avedikian. In *Chemistry and Physics of Aqueous Gas Solutions* (ed. W. A. Adams). Electrochemical Society 1975, pp. 172–82.
160. N. Desrosiers & J. E. Desnoyers. *Can. J. Chem.* **54** (1976), 3800–08.
161. J. Lara, G. Perron & J. E. Desnoyers. *J. Phys. Chem.* **85** (1981), 1600–05.
162. M. Palma & J. P. Morel. *J. Solution Chem.* **8** (1979), 767–77.
163. M. Palma & J. P. Morel. *Can. J. Chem.* **59** (1981), 3248–51.
164. G. Roux-Desgranges, A. H. Roux, J. P. Grolier & A. Viallard. *J. Colloid Interface Sci.* **84** (1981), 536–45.
165. A. Roux, G. Roux-Desgranges, J. P. Grolier & A. Viallard. *J. Colloid Interface Sci.* **84** (1981), 250–62.
166. G. Scatchard. *Chem. Rev.* **44** (1949), 7–35.